Pablo Koch Medina
Sandro Merino

Mathematical Finance and Probability

A Discrete Introduction

Birkhäuser Verlag
Basel · Boston · Berlin

Authors

Pablo Koch Medina
Swiss Re
Mythenquai 50/60
8022 Zürich
Switzerland
e-mail: pablo_kochmedina@swissre.com

Sandro Merino
UBS AG
Bahnhofstrasse 45
8098 Zürich
Switzerland
e-mail: sandro.merino@ubs.com

2000 Mathematics Subject Classification 60-01; 91-01, 91B28, 91B30

A CIP catalogue record for this book is available from the
Library of Congress, Washington D.C., USA

Bibliographic information published by Die Deutsche Bibliothek
Die Deutsche Bibliothek lists this publication in the Deutsche Nationalbibliografie; detailed bibliographic data is available
in the Internet at <http://dnb.ddb.de>.

ISBN 3-7643-6921-3 Birkhäuser Verlag, Basel – Boston – Berlin

© 2003 Birkhäuser Verlag, Postfach 133, CH-4010 Basel, Switzerland
Member of the BertelsmannSpringer Publishing Group
Cover design: Micha Lotrovsky, CH-4106 Therwil, Switzerland
Printed on acid-free paper produced from chlorine-free pulp. TCF ∞
Printed in Germany
ISBN 3-7643-6921-3

9 8 7 6 5 4 3 2 1

To Moira
To Pablo and Matilde

Contents

You can fool some people some time but you can't fool all the people all the time.
                                                      Bob Marley

Chapter 1

Introduction

On what grounds can one reasonably expect that a complex financial contract solving a complex real-world issue does not deserve the same thorough scientific treatment as an aeroplane wing or a micro-processor? Only ignorance would suggest such an idea.

E. Briys and F. De Varenne

The objective of this book is to give a self-contained presentation of that part of mathematical finance devoted to the pricing of derivative instruments. During the past two decades the pricing of financial derivatives — or more generally: mathematical finance — has steadily won in importance both within the financial services industry and within the academic world. The complexity of the mathematics needed to master derivatives techniques naturally resulted in a high demand for quantitatively oriented professionals (mostly mathematicians and physicists) in the banking and insurance world. This in turn triggered a demand for university courses on the relevant topics and at the same time confronted the mathematical community with an interesting field of application for many techniques that had originally been developed for other purposes. Most probably this development was accelerated by an ever more applied orientation of the mathematics curriculum and the fact that finance institutions were often willing to generously support research in this field.

The objective and the intended audience

The material presented in this book will be accessible to someone having a working knowledge of linear algebra and calculus as taught in the first three semesters at most European universities. All additional material is developed from the very beginning as needed. In particular the book also offers an introduction to modern probability theory, albeit mostly within the context of finite sample spaces.

The style of presentation we have chosen will appeal to

- financial economics students seeking an elementary but rigorous introduction to the subject;

- mathematics and physics students from the third semester onwards who would like to become acquainted with a modern applied topic such as this; and

- mathematicians, physicists or quantitatively inclined economists working in the financial industry who have not previously had a deep exposure to probability theory, but wish to understand in a rigorous manner the main ideas behind option pricing.

We have strived for clarity rather than conciseness and have opted for redundancy where deemed appropriate. In particular we have often chosen to repeat the structure of presentation when introducing new complexity to the models. While to some this may appear to be unnecessarily tedious, our experience has been that it helps get a thorough and more lasting understanding of the underlying principles. In the subject matter of this book finance and the rigid logic imposed by the use of mathematics blend, uncovering many financial phenomena whose existence would otherwise remain hidden from the naked eye. Although we approach our subject with the eyes of a mathematician — emphasizing the underlying structure and providing full proofs of almost all results — we have tried to stress the financial interpretation of the topics covered. Hence, this book can also be seen as a book on finance for mathematicians. Moreover, we believe that presenting topics from probability, stochastic processes and linear algebra using essentially one field of application for their illustration provides the reader with a rich source of intuition helping them get a firmer grasp of the mathematical concepts.

The basic theme

Our basic theme is the study of prices in *securities markets* in an uncertain environment. There, agents exchange securities for money, a *security* being a contract entitling the holder to a pattern of future payments contingent on the state in which the world turns out to be at the date they are due. Point of departure will be the existence of a finite set of *basic* traded securities selling for a price determined by market mechanisms, i.e. by supply and demand. We will be content to take prices of these securities as exogeneously given "observables": We will not seek to explain how a consensus about these prices is reached among the economic agents in the market. This too is an interesting and important question, but one whose treatment would require us to go into equilibrium theory, distracting us from our main concern which will be the following question: whether or not *contingent claims*, i.e. arbitrary patterns of payments contingent on the state of the world, can be "produced" by setting up a suitable (dynamic) portfolio of basic instruments.

The process of setting up a dynamic portfolio mimicking the payments of the contingent claim is called *replication*. If a contingent claim can be replicated, i.e. if it is *attainable*, we can then effectively "produce" a new security entitling the holder to that particular pattern of payments. The price of the newly created security will of course be the cost of carrying out the replicating strategy. These newly created securities are also called *derivative securities* or just *derivatives*, because they are "derived" from the basic market instruments.

The activity of constructing new securities from the basic ones is sometimes referred to as *financial engineering*. This seems an adequate label for this occupation, since we are not really concerned with a fundamental explanation of the economic environment; we just take the market for basic securities as exogenously given and their prices and explore what we may build from there on. The only thing we will require for our elementary market is that no *arbitrage opportunities* exist, i.e. that there are no opportunities for potential gains (with no potential losses) at zero-cost. Financial engineering is closely related to *financial risk management*. With an increasing awareness of the many financial risks the capital of a given company is exposed to (interest rate risk, stock price risk, currency risk, credit risk, etc.), methods have been developed to effectively manage them. Managing these risks means consciously choosing which risks to be exposed to and to what degree. This is sometimes referred to as *taking a position* in the different risks. Financial engineering provides the tools for a systematic and efficient management of financial risks.

What is mathematical modelling?

A few words on the nature of mathematical modelling. When trying to apply mathematical methods to a particular problem the first step will consist in giving a mathematical description of the environment in which the problem is embedded, i.e. in proposing a *mathematical model*. The problem which is being addressed will then have to be formulated in terms of mathematical relations between the objects defined in the model. Using mathematical reasoning, a solution will hopefully be found. Generally, the main object in constructing a model is to be able to predict the future behavior of what is being modelled, to explain observed phenomena in terms of a minimal set of assumptions, or to uncover new ones that may lie unrecognized, camouflaged by the rich and often confusing detail of reality. While *deterministic* models will deliver unambiguous predictions, *probabilistic* ones will essentially provide an assignment of probabilities for all possible outcomes. Which type of model to choose will naturally depend on the particular issue under consideration.

Ideally, the model should specify all factors which have an influence on the issue at hand and their interaction. However, insufficient understanding or tractability considerations will force us to settle for a less accurate translation of reality into a mathematical setting: a good deal of information will be sacrificed during the modelling process. Thus, successful modelling will depend on the ability to select an "appropriate" set of explanatory factors and to specify mathematical relations between them. Here "appropriate" is to be understood as relevant for the particular use. In any case, every model is by the very nature of its genesis incomplete and transitory in character. Its success will depend on its usefulness, i.e. its cognitive and predictive power, which can only be assessed through testing.

What are the elements of any mathematical model of a securities market?

What should be specified when modelling a market where a given finite number of securities are traded? We emphasize that we will remain within the context of financial engineering. We do not address the issue of how economic agents make their allocation choices. This will mean that we need to make very few assumptions on the behavior of individuals. In fact, we will make only the following one:

- In any state of the world they will prefer to have more wealth than less.

Usually, when individual preferences are modelled, it has to be specified for any two goods traded in the market, whether the individual prefers one to the other, or is indifferent between the two of them. We will not make any assumptions on this whatsoever. If our individual has to choose between two *comparable* [1] patterns of payments, he will choose the one with the higher payments. But if the patterns are not comparable, we cannot say which of them he will prefer.

We now list the items a model should specify:

- *Time horizon*: This entails specifying a date T at which all economic activity stops.

- *Trading dates*: Here, dates between today (t=0) and the time horizon T are specified, where trading is assumed to take place.

- *States of the world*: This requires listing all possible states the world can be in at the trading dates; states which are relevant to the economic environment we desire to model.

- *Probabilities*: Probabilities should be specified for the possible states of the world.

- *Traded securities*: This requires the specification of the behavior (contingent on the states of the world) of prices of the securities available in the economy we seek to model.

---

[1] By *comparable* payments, we mean that in all states of the world, one of them pays at least as much as the other. They would not be comparable if in some states one of them pays more than the other, while paying less than the other in other states.

- *Trading rules*: How are securities traded? Are there transaction costs? Do taxes have to be paid? Can one buy only a whole number of units of a particular security? Are there any restrictions on borrowing?

A few remarks are necessary:

- The time horizon can be finite or infinite. We shall consider mostly a finite time horizon.

- Models with a continuum of trading dates are said to be *continuous time models*. Models where trading dates form an infinite but discrete set are called *discrete time models*. The name *finite time models* is reserved for models with a finite set of trading dates. In these notes we shall deal exclusively with this case.

- In finite time models we distinguish between *single-period* and *multi-period* models. In the former kind of model there are two relevant dates $t = 0$ and $t = T$, the only relevant period is $[0, T]$. In the latter more than one date is given $0 \stackrel{\text{def}}{=} t_0 < t_1 < t_2 < \ldots < t_{n-1} < t_n \stackrel{\text{def}}{=} T$. The relevant periods are $[t_0, t_1], [t_1, t_2], \ldots, [t_{n-2}, t_{n-1}], [t_{n-1}, t_n]$.

- The specification of the possible states of the world is a crucial element. In a sense we have to isolate the problem, identifying the economic variables relevant to it.

What we aim at

Having introduced some of the issues we will address here, we shall now return to a description of the objective of this book. We have endeavored to provide, within a simple context, a sound understanding of some of the most important issues pervading modern mathematical finance. In particular we seek to explain and clarify the concept of a change of measure, the cornerstone of the so-called martingale approach to pricing contingent claims. The reader will thus become acquainted with the major ideas contained in the seminal papers of J.M. Harrison and D.M. Kreps (1979) and Harrison and S.R. Pliska (1981). The topic is by its very nature rather technical. In particular, the continuous-time case requires a deep understanding — or, at the very least, a well-developed intuition — of a variety of advanced topics in modern probability theory (stochastic integrals, Ito's formula, etc.). The finite-time case treated in this book, however, may be approached resorting to a great deal less of mathematics. This allows a focus on modelling and interpretation issues. Moreover, the finite-time case can be cast in a language similar to that of continuous-time finance rendering the transition to the latter theory more amenable.

The concept for the presentation of the material has been successfully tested in courses at the University of Berne (Switzerland) in 1996, by the first author, and at the University of Strathclyde (Glasgow, United Kingdom) in 1998, by the second author. Moreover, both authors have used parts of the book in various internal courses at Winterthur Insurance Company, Swiss Re, and UBS.

Alternative ways to read this book

As stated above the desire to have an essentially self-contained presentation of the material forced us to include material which some of the readers will already know. This refers particularly to some of the more elementary probability theory. However, we have made an effort to strictly separate "mathematical" from "financial" topics, so that the reader can easily leave out those topics which he or she is already masters without interrupting the flow of ideas. The financial topics are included in chapters 2, 6, 9, 11, 12, 14, and 16.

Acknowledgements

The first author would like to thank Fabio Trojani for many a long discussion they had during 1996–1997 on how a financial economist, albeit one with a strong mathematical focus, looks at many of these topics. The first author is also indebted to Swiss Re for providing the necessary support to finish this book. We would also like to thank the following individuals for proofreading parts of the manuscript: Niklaus Bühlmann, Heiner Schwarte, Frank Weber, and Andreas Wyler.

Chapter 2

A Short Primer on Finance

One of the major advances in financial economics in the past two decades has been to clarify and formalize the concept of "no arbitrage" and to apply this idea systematically to uncover hidden relationships in asset prices.

H.R. Varian

In this first chapter we address most of the basic issues of our subject within the context of an extremely simple example. The mathematics involved do not go beyond solving a system of two linear equations with two unknowns. However simple, this example will serve as a vehicle to illustrate the main ideas.

2.1 A One-Period Model with Two States and Two Securities

We begin by describing the framework we will be working in. This involves formulating the investor's financial engineering problem, introducing the market for securities and specifying step by step the mathematical objects that are used to model the underlying economic ideas. We start with a rather conversant description of the sort of real life situation we would like to model and then proceed to give a more rigorous mathematical formulation.

2.1.1 The Investor's Financial Engineering Problem

The main character in our play is an investor who lives in an uncertain economic environment, i.e. an economy whose future state we cannot know for certain. He has a time horizon of say one month, meaning that his basic problem is that of transferring his wealth from today $(t = 0)$ to next month $(t = 1)$ in a way that befits his needs.

Transferring his wealth from today to tomorrow entails entering a *financial contract* with some counterparty. Such a contract specifies for each possible future state of the economy a payment which he will either receive from or have to pay to the counterparty at time $t = 1$. The set of all conceivable financial contracts is sometimes called the set of *alternatives*. Thus, an alternative is a potential financial contract.

Sometimes we will refer to the pattern of payments generated by a given alternative as its *payoff* or *pay-out*.

The target alternative

In our context the needs of our investor will be captured by the specification of a *target alternative*. In other words, he will specify his needs by stating how much wealth he wishes to possess at time $t = 1$ in each of the states the economy might be in at that time. If he is able to enter a contract entitling him to the desired target alternative, at time $t = 1$, when the true state of the economy has crystalized, he will make or receive exactly the payment he had programmed for that particular state.

Financial claims

Among financial contracts we would like to single out those that bear an obligation to make a payment for only one of the contracting parties. The party obliged to make a payment is called the *issuer*, the one entitled to receive the payment the *holder*. This type of financial contract will be called a *financial claim*, since the holder has a claim on the issuer. From the point of view of the issuer, a claim is a way of obtaining funds today in exchange for the promise to pay tomorrow a given amount which may be contingent on the state the economy turns out to be in. From the point of view of the holder, a claim is an investment which enables the transfer of wealth today into the future. Sometimes to stress the fact that the payment may depend on the state of the world at time $t = 1$ we use the term *contingent claim*. Claims which pay a fixed amount no matter which state the world turns out to be in are called *deterministic*.

The market for securities

We shall be concerned with an economy where there is a market in which financial claims are traded. However, not every claim which may be conceived will also be traded in the market. Those that are traded will be called *financial securities*, or *securities* for short. We shall assume that financial securities are claims that pay a (strictly) positive amount in each state of the economy.

The prices at which financial securities trade will be called *market prices*. We shall assume that market prices are *equilibrium prices*. Offer and demand will drive prices: they will adjust until offer and demand match, i.e. the market will have cleared so that at prevailing prices none of the market participants has an incentive to trade any longer. We emphasize that we will not be concerned with modelling the mechanism we just sketched by which the market arrives at these prices. Market prices will be *exogeneous variables* in our model.

Replication: the key to financial engineering

The way our investor will operate is the following. He will specify a desired target alternative. By buying and selling securities at market prices, he will then try to set up a portfolio in such a way that the pattern of payments generated by the portfolio as a whole at time $t = 1$ is identical to his target alternative. Any such portfolio will be called a *replicating portfolio* for that alternative. What our investor is in fact doing is attempting to "engineer" or "produce" an alternative which suits his needs by using the market for securities as a provider of raw materials. This is why the activity of structuring financial contracts is sometimes referred to as *financial engineering*.

Replication: the key to valuation

Once our investor knows how to replicate alternatives he will also know how to put a value on them. He will know the cost of setting up a replicating portfolio and thus the "production costs". This will lead us to the notion of *fair value*. In fact what we have just said gives away the key idea of the theory of valuation of financial contracts: replicate the contract by a portfolio of traded securities and value it at the cost of replication. This means that the valuation of financial contracts is a "relative" valuation theory: it takes financial contracts and values them relative to the value of traded securities.

At this point it may be worthwhile making a remark about terminology. It is good discipline to make a distinction between price and value of financial contracts. Prices generally refer to sums of money for which a contract is transacted. Fair value refers to the value a contract should have given market prices for the basic securities. Following standard practice however, we will not always stick to this distinction an whenever we write price we will either mean the price of a basic security or the fair value of a financial contract (which for a basic security coincides with its price).

Two issues we will not address

We already mentioned that we will not address the question of how markets arrive at equilibrium prices. We will take market prices as given.

Another issue which we will not contemplate in this book is how our investor arrives at the conclusion that a particular target alternative suits his needs best. We will only address the issue of how, after having chosen his target alternative, he may go about generating it by setting up a suitable replicating portfolio.

2.1.2 The Uncertain Economy: A Two-State Model

As we already noted, an uncertain economy is an economy whose present state — its state at time $t = 0$ — is known, but whose future state — its state at time $t = 1$ — can be any one of a given set of possible states. Which of the possible states it actually turns out to be will only become apparent at time $t = 1$. We will sometimes say that the (*true*) state of the economy is *revealed* at time $t = 1$. Of course, at that time we will know for certain which state the economy is in, but from the point of view of time $t = 0$ the future state of the world is uncertain.

In this first chapter we focus on an economy which at time $t = 1$ can only assume one of two possible states. Somewhat arbitrarily — but suggestively — we will call them the "good" and the "bad" state, respectively. Sometimes, for stylistic reasons, we will use the term "state of the world" instead of "state of the economy". The possible future states of the economy, can be represented by the elements of the set

$$\Omega \stackrel{\text{def}}{=} \{g, b\} \,,$$

where, naturally, g stands for "good" and b for "bad". The set Ω is also referred to as the underlying *sample space*, thus emphasizing the random character of the future states.

2.1.3 Probabilities

The economy we are considering is uncertain and modelled by the underlying sample space of all possible outcomes, which in our case is given by the set $\Omega = \{g, b\}$. Both states will have a chance of occurring with a certain probability. Let $p \in (0, 1)$ denote the probability that at time $t = 1$ the world will be in the "good" state. It follows that it will be in the "bad" state with probability $1 - p$. We may define for each subset A of Ω its probability of occurring by setting

$$P(A) \stackrel{\text{def}}{=} \begin{cases} 0 & \text{if } A = \emptyset, \\ p & \text{if } A = \{g\}, \\ 1 - p & \text{if } A = \{b\}, \\ 1 & \text{if } A = \Omega. \end{cases}$$

We call P the *underlying probability measure (or distribution)* [1].

When modelling the real world it may be an extremely controversial task to determine the underlying probability distribution. Luckily, as will become apparent later on, for the applications we have in mind we do not have to model P exactly: It is not necessary to actually "know" P. We need only assume that when modelling the market we know which events are possible (in our case g and b).

---

[1] In these book we shall use the terms "probability distribution" and "probability measure" interchangeably.

This may sound surprising right now, but it is not if we recall that we are not interested in forecasting where the market will be in the future. The task we are confronted with is the replication of the target alternative of our investor. He wants to set up a portfolio which gives him in each of the possible states a pre-specified amount, regardless of the probability with which each of the states might occur. Probabilities may, however, play a role when deciding on which target alternative to choose, but remember: this is not one of the issues we will be addressing!

2.1.4   Financial Contracts and Random Variables

An alternative, a candidate for a financial contract, can be specified by stating its payoff, i.e. the amount of money to be paid or received at time $t = 1$. It can therefore be represented by a random variable on the underlying sample space Ω, i.e. by a function $X : \Omega \to \mathbb{R}$. Thus, $X(g)$ and $X(b)$ are the amounts due should the economy wind up being in the "good" and the "bad" states, respectively. Given a state $\omega \in \Omega$, positivity of $X(\omega)$ will mean a payment to be received, while negativity will represent a payment to be made. We denote the set of alternatives by $\mathcal{A}$.

The vector space of alternatives

For two alternatives $X, Y \in \mathcal{A}$ we may define their sum $X + Y$ by

$$(X + Y)(\omega) \stackrel{\text{def}}{=} X(\omega) + Y(\omega)$$

and the product of X and a scalar $\lambda \in \mathbb{R}$ by

$$(\lambda X)(\omega) \stackrel{\text{def}}{=} \lambda X(\omega)$$

for each $\omega \in \Omega$. Together with these operations the set of alternatives $\mathcal{A}$ is a vector space.

Random variables as vectors

Every random variable $X : \Omega \to \mathbb{R}$ induces the vector $(X(g), X(b)) \in \mathbb{R}^2$. On the other hand, given an arbitrary vector $(x_g, x_b) \in \mathbb{R}^2$ we may define a unique random variable by setting $X(g) \stackrel{\text{def}}{=} x_g$ and $X(b) \stackrel{\text{def}}{=} x_b$.

Thus, random variables on Ω can be identified with vectors in $\mathbb{R}^2$. Moreover, with this identification, the addition and multiplication by a scalar for random variables defined above corresponds exactly to the usual addition and multiplication by a scalar on $\mathbb{R}^2$. We may therefore even identify $\mathcal{A}$ with $\mathbb{R}^2$ as vector spaces without any risk of confusion. We will use these alternative views at our discretion, choosing whichever is more convenient for a particular situation.

2.1.5   Deterministic and Contingent Alternatives

Recall that alternatives that pay a fixed amount at maturity regardless of the state of the world are called deterministic. A claim $X : \Omega \to \mathbb{R}$ is thus deterministic if there exists a constant $C \geq 0$ such that

$$X(g) = X(b) = C \ .$$

Deterministic alternatives represent certain payments. Genuine contingent alternatives are alternatives $X : \Omega \to \mathbb{R}$ such that $X(g) \neq X(b)$. Contingent alternatives reflect true riskiness.

2.1.6   Claims and Positive Random Variables

A claim was defined to be an alternative which entitles the holder to the receipt of payments. Therefore, claims correspond to *positive random variables*: random variables $X : \Omega \to \mathbb{R}$ which assume only non-negative values, i.e. for which

$$X(g), X(b) \geq 0 \ .$$

Positive random variables for which either $X(g) > 0$ or $X(b) > 0$ holds are called *strictly positive*. If they satisfy $X(g) > 0$ *and* $X(b) > 0$ they are called *strongly positive*.

Remember that we have assumed that securities are claims which entitle the holder to a positive payment in each state of the economy. They therefore correspond to strongly positive random variables.

2.1.7   The Market for Securities: A Two Securities Model

In our simple economy only two securities will be traded in the market: the *risk-free* and the *risky* security, respectively.

- The risk-free security, or the *zero-bond*, is characterized by the fact that it entitles the holder to receive a payment of 100 currency units at time $t = 1$, regardless of the state the world turns out to be in. The amount promised to be paid at the end of the period is called the *face value* or *nominal value* of the zero-bond. The time at which the face value is repaid is called the *maturity (date)*. In our case, of course, the maturity is $t = 1$. If our investor buys the zero-bond he will know with certainty the amount of money he will receive at time $t = 1$. Hence the name "risk-free security".

  The representation of this security as a random variable $B_1 : \Omega \to \mathbb{R}$ is of course
  $$B_1(g) = B_1(b) = 100 \ .$$

  Viewed as a vector we have
  $$B_1 = (100, 100) \ .$$

- The risky security, or the *stock*, is characterized by the fact that it entitles the holder to different (strictly positive) payments depending on whether the world turns out to be in the "good" or in the "bad" state, respectively. If our investor buys the stock he will not know with certainty which payment he will be entitled to at time $t = 1$. This explains the name "risky" security.

  This security can be represented by a strictly positive random variable $S_1 : \Omega \to \mathbb{R}$ which can be written as

  $$(S_1(g), S_1(b))$$

  when viewed as a vector. Recall that strictly positive means that $S_1(g) > 0$ and $S_1(b) > 0$ both hold.

In the market we are considering, only these two securities are traded. The participants in this market (to which our investor belongs) will be called *traders* or *investors*. We shall make the following assumptions:

- We shall denote the price of the zero-bond by B_0 and that of the stock by S_0. Both of these prices are assumed to be strictly positive, i.e.

  $$B_0 > 0 \quad \text{and} \quad S_0 > 0 .$$

  This is because, since both securities entitle the holder at time $t = 1$ to potential benefits without any obligations, they should have a positive value to the buyer.

- The zero-bond and the stock are both *infinitely divisible*, i.e. they can be bought or sold in any quantity. For example one could buy $\sqrt{2}$ units of the stock.

- The market is *frictionless*, i.e. there are no transaction costs or taxes to be paid when a security changes hands.

- At time $t = 0$, each of these traders is entitled to buy or sell securities.

- We allow short-sales of the stock to be carried out. Here, a *short-sale* of the stock entails selling the stock while not owning it. For that to be feasible the trader has to be able to borrow the stock at time $t = 0$ in order to perform the sale. At time $t = 1$ the trader will buy it at the then prevailing price in order to give it back.[2]

---

[2]Short-selling the stock is a strategy often used to speculate on the price of the stock falling: sell the stock short for S_0 and invest the proceeds in the zero-bond. If the price of the zero-bond is less than its face value you will obtain more than S_0 at time $t = 1$. If at that time the price of the stock has fallen you will have to replace, to its original owner, a stock now worth less than S_0. This will leave you with a profit. Of course, if the stock price increases above the amount you obtain from the zero-bond position you will suffer a loss. Therefore, short-selling generally has a speculative character. We will see later that it may be riskless if the market admits so-called arbitrage opportunities.

- We also allow *borrowing* money at fixed conditions, which is equivalent to a short-sale of the zero-bond. The proceeds of the short-sale correspond to the borrowed amount at time $t = 0$. At time $t = 1$ we have to give the zero-bond back, i.e. we have to pay the fixed amount 100 currency units.

- At time $t = 1$ traders will liquidate their positions, i.e. receive payments from the claims they sold or make payments if they have borrowed money or sold short stock[3].

2.1.8   The Price Processes for the Securities

Recall that S_1 and B_1 were defined as random variables from Ω into $\mathbb{R}$, while S_0 and B_0 were positive scalar numbers. It is convenient to also interpret B_0 and S_0 as random variables from Ω into $\mathbb{R}$. This can be done in a very natural way by setting:

$$B_0(g) \overset{\text{def}}{=} B_0(b) \overset{\text{def}}{=} B_0$$
$$S_0(g) \overset{\text{def}}{=} S_0(b) \overset{\text{def}}{=} S_0.$$

We may now use use the following self-explanatory tables in order to better visualize the price process of the bond and the stock:

B_t	g	b
$t = 0$	B_0	B_0
$t = 1$	100	100

S_t	g	b
$t = 0$	S_0	S_0
$t = 1$	$S_1(g)$	$S_1(b)$

For example

B_t	g	b
$t = 0$	90.90	90.90
$t = 1$	100	100

S_t	g	b
$t = 0$	100	100
$t = 1$	111	103

describes a market where the zero-bond with a face value of 100 currency units sells at 90.90, while the stock — paying at time $t = 1$ the amount of 111 currency units if the economy turns out to be in the "good" state and 103 currency units if it ends up in the "bad" state — sells at 100.

---

[3]Note that, as in the case of short-selling a zero-bond, a short-sale of the stock is also essentially like borrowing at time $t = 0$ the proceeds of the short-sale. The only difference from borrowing money at fixed conditions is that now we do not know exactly how much we have to pay back at time $t = 1$. It might be $S_1(g)$ or $S_1(b)$ depending on the state the economy is revealed to be in at time $t = 1$.

2.1.9   Return of the Zero-Bond

The *return* of the zero-bond is defined as

$$r \stackrel{\text{def}}{=} \frac{B_1 - B_0}{B_0} = \frac{100 - B_0}{B_0}.$$

Since the price of the zero-bond B_0 is strictly positive the return is well defined and is easily seen to satisfy $-1 < r$.
We may rewrite the price of the zero-bond as

$$B_0 = \frac{B_1}{1+r} = \frac{100}{1+r}. \tag{2.1}$$

In fact, due to this relationship, instead of quoting the price of the zero-bond directly we may do so by specifying its return r.
Selling the bond for B_0 is essentially borrowing B_0 at time $t = 0$ and returning

$$B_1 = 100 = B_0 \cdot (1+r) = B_0 + r \cdot B_0 \tag{2.2}$$

i.e. returning the borrowed amount B_0 and the interest $r \cdot B_0$. For this reason r is sometimes called the *risk-free interest rate*, or the *risk-free rate* for short. As mentioned before, we may quote the price of the zero-bond by quoting the risk-free rate. For instance, if we say that the risk-free rate is $r = 10\%$, then the features of the zero-bond will be given by the following table:

B_t	g	b
$t = 0$	90.90	90.90
$t = 1$	100	100

We will require that r be non-negative. If $r = 0$, then $B_0 = 100$. Finally, if $r > 0$ we have $B_0 < 100$, meaning that the bond will *appreciate*.

Non-negative rates

In economics one usually distinguishes between *nominal* and *real* interest rates. The nominal interest is just the risk-free rate we introduced above. The real rate, however, is based on the purchasing power of the amount received at maturity, i.e. it will correspond to the risk-free rate less the rate of inflation in the economy.
It makes economic sense to assume that nominal rates are non-negative. Were the risk-free rate to be negative, our investor could always opt for keeping his money in his pocket instead of investing in the bond. By doing that he would be virtually mimicking what amounts to holding a zero-bond with zero return. Hence, nobody would have any incentive to buy a zero-bond if the risk-free rate were to be negative.
Thus, real rates are the only types of negative rates which make economical sense. Since in this book we are only concerned with nominal rates we preclude r from being negative, although allowing it would represent no additional complexity.

2.1.10   Return of the Stock

At time $t = 1$ the stock will have a different payoff $S_1(g)$ and $S_1(b)$, depending on the state of the economy turning out to be "good" or "bad", respectively. Therefore, the *return* of the stock will also depend on which state the economy is in at time $t = 1$. It will be

$$y(g) \overset{\text{def}}{=} \frac{S_1(g) - S_0}{S_0}$$

if the state of the world is "good" and

$$y(b) \overset{\text{def}}{=} \frac{S_1(b) - S_0}{S_0}$$

if the state of the world is "bad".

As for the return of the zero-bond, it is easy to see that $-1 < y(g)$ and $-1 < y(b)$ both hold.

Since the return of the stock is in fact a random variable $y : \Omega \to \mathbb{R}$, we cannot rewrite its price as we did for the bond in equation (2.1). However, we can find an expression linking the payoff at time $t = 1$, the price at time $t = 0$ and the return similar to equation (2.2):

$$S_1(g) = S_0 \cdot (1 + y(g)) \qquad \text{and} \qquad S_1(b) = S_0 \cdot (1 + y(b)). \qquad (2.3)$$

We will always assume that

$$y(b) < y(g) ,$$

i.e. in the "good" state the stock has a return which is higher than in the "bad" state. This assumption will spare us the formal necessity of having to distinguish between the cases $y(g) < y(b)$ and $y(b) < y(g)$.

Moreover, for the same reason for which we assumed that the risk-free rate was non-negative, we will assume that $y(g) \geq 0$. Otherwise, keeping the money in one's own pocket would be more attractive than buying the stock.

If $y(b) > 0$, then $S_1(b) = S_0 \cdot (1 + y(b))$ implies that the stock will appreciate if the state of the world is "bad". If $y_b = 0$ it means that the price will remain the same and if $y(b) < 0$ that the stock will depreciate. Of course, the analogous statements hold for the state of the world "good" (except that by assumption we do not allow the stock to depreciate in the "good" state).

Note that the price process of the stock is completely characterized by the initial price S_0 and the random variable $y : \Omega \to \mathbb{R}$ describing the return, thus by the triple $(S_0, y(g), y(b))$. For instance the features of the stock price process as specified by $S_0 = 100$, $y(g) = 11\%$ and $y(b) = 3\%$ are given by the table

S_t	g	b
$t = 0$	100	100
$t = 1$	111	103

2.1.11 Markets and Price Systems

We have seen that we may completely specify the market within our model by choosing the *market parameters* r, S_0, $y(g)$ and $y(b)$. For convenience we call the set of parameters

$$\{r, S_0, y(g), y(b)\}$$

a *price system* if the following three conditions, which we assume from now on, are satisfied:

- $0 \leq r$;

- $S_0 > 0$;

- $0 \leq y(g)$;

- $-1 < y(b) < y(g)$.

These conditions merely summarize the conditions we have imposed on our economy up to now.

An example for a price system is $(r, S_0, y_g, y_b) = (10\%, 100, 11\%, 3\%)$ implies the following tables for the price processes of the zero-bond and the stock, respectively:

B_t	g	b
$t = 0$	90.90	90.90
$t = 1$	100	100

S_t	g	b
$t = 0$	100	100
$t = 1$	111	103

2.1.12 Portfolios

For a trader the result of trading is a *portfolio*, i.e. it is a combination of an amount of bonds and an amount of stock. A portfolio can be represented by a pair

$$(\alpha, \beta) \in \mathbb{R}^2 ,$$

where α and β denote the number of units of bonds and of stock in the portfolio, respectively. Recall that we have allowed short-selling and borrowing and that both securities are infinitely divisible. Hence, both α and β can be arbitrary numbers, positive — representing holding the security or a so called *long position* — or negative — representing a short-sale of the security or a *short position*.
Consider the following examples of portfolios:

a) The portfolio $(1, 1)$ contains one unit of the bond and one unit of the stock.

b) The portfolio $(\sqrt{2}, -1)$ represents $\sqrt{2}$ units of the bond and the short-sale of one unit of the stock.

c) The portfolio $(-1, 1)$ entails having borrowed 100 currency units at the risk-free rate r and having bought one unit of the stock.

2.1.13   The Initial Value of a Portfolio

Assume that we are given a portfolio $(\alpha, \beta) \in \mathbb{R}^2$. We define its *initial value* by

$$V_0^{(\alpha,\beta)} \stackrel{\text{def}}{=} \alpha \cdot B_0 + \beta \cdot S_0 \ .$$

We thus identify the initial value of (α, β) as the cost[4] of setting up this portfolio at time $t = 0$.

From the definition of the initial value of a portfolio it is immediate that the mapping assigning to each portfolio $(\alpha, \beta) \in \mathbb{R}^2$ its initial value $V_0^{(\alpha,\beta)}$ is linear, i.e.

$$V_0^{(\alpha,\beta)+(\tilde{\alpha},\tilde{\beta})} = V_0^{(\alpha,\beta)} + V_0^{(\tilde{\alpha},\tilde{\beta})},$$
$$V_0^{\lambda(\alpha,\beta)} = \lambda V_0^{(\alpha,\beta)}.$$

As in the case of B_0 and S_0 it is sometimes convenient to view $V_0^{(\alpha,\beta)}$ as a random variable on Ω by setting

$$V_0^{(\alpha,\beta)}(g) \stackrel{\text{def}}{=} V_0^{(\alpha,\beta)}(b) \stackrel{\text{def}}{=} V_0^{(\alpha,\beta)} \ .$$

2.1.14   The Terminal Value of a Portfolio

The *terminal value* of a portfolio $(\alpha, \beta) \in \mathbb{R}^2$ is defined as

$$V_1^{(\alpha,\beta)}(g) \stackrel{\text{def}}{=} \alpha B_1(g) + \beta S_1(g)$$

if the state of the world is "good" and

$$V_1^{(\alpha,\beta)}(b) \stackrel{\text{def}}{=} \alpha B_1(b) + \beta S_1(b)$$

if the state of the world is "bad". Observe, that we have

$$V_1^{(\alpha,\beta)}(g) = \alpha B_1(g) + \beta S_1(g) = \alpha 100 + \beta S_0(1 + y(g))$$

and

$$V_1^{(\alpha,\beta)}(b) = \alpha B_1(b) + \beta S_1(b) = \alpha 100 + \beta S_0(1 + y(b)) \ .$$

The *terminal value* of (α, β), i.e. its value at time $t = 1$ represents what the holder will obtain when liquidating the portfolio at the end of the period depending on the state of the economy prevailing at that time.

It is immediate from the definition of terminal value that assigning to each portfolio $(\alpha, \beta) \in \mathbb{R}^2$ its terminal value $V_1^{(\alpha,\beta)} \in \mathcal{A}$ is a linear operation, i.e.

$$V_1^{(\alpha,\beta)+(\tilde{\alpha},\tilde{\beta})} = V_1^{(\alpha,\beta)} + V_1^{(\tilde{\alpha},\tilde{\beta})},$$
$$V_1^{\lambda(\alpha,\beta)} = \lambda V_1^{(\alpha,\beta)}.$$

---

[4]Remember, we have a frictionless market, so that there are no transaction costs.

2.1.15   The Value Process of a Portfolio

For each portfolio we have defined its initial and terminal values, i.e. we have described its *value process*.

Of course, as in the case of the price processes of securities, we may also use a table for visualizing the value process of a portfolio:

$V_t^{(\alpha,\beta)}$	g	b
$t=0$	$\alpha\frac{100}{1+r} + \beta S_0$	$\alpha\frac{100}{1+r} + \beta S_0$
$t=1$	$\alpha 100 + \beta S_0(1 + y(g))$	$\alpha 100 + \beta S_0(1 + y(b))$

For example the price system $(r, S_0, y_g, y_b) = (5\%, 100, 10\%, -10\%)$ implies the following prices

B_t	g	b
$t=0$	95.24	95.24
$t=1$	100	100

S_t	g	b
$t=0$	100	100
$t=1$	110	90

In this situation the portfolios $(\alpha, \beta) = (3, 2)$ and $(\tilde{\alpha}, \tilde{\beta}) = (-3, 5)$ have the following respective value tables

$V_t^{(\alpha,\beta)}$	g	b
$t=0$	485.72	485.72
$t=1$	520	480

$V_t^{(\tilde{\alpha},\tilde{\beta})}$	g	b
$t=0$	214.28	214.28
$t=1$	250	150

2.1.16   The Return of a Portfolio

The *return* of a portfolio $(\alpha, \beta) \in \mathbb{R}^2$ is defined as

$$r_{(\alpha,\beta)} \overset{\text{def}}{=} \frac{V_1^{(\alpha,\beta)} - V_0^{(\alpha,\beta)}}{V_0^{(\alpha,\beta)}} \, .$$

Since the terminal value $V_1^{(\alpha,\beta)}$ of the portfolio (α, β) generally depends on the state of the world at time $t = 1$ (unless $\beta = 0$), the return of a portfolio is a random variable. We may write

$$V_1^{(\alpha,\beta)} = V_0^{(\alpha,\beta)} \cdot (1 + r_{(\alpha,\beta)}) \, .$$

It is not surprising and it is easy to see that

$$r_{(\alpha,\beta)} = \frac{\alpha \cdot B_0}{\alpha \cdot B_0 + \beta \cdot S_0} \cdot r + \frac{\beta \cdot S_0}{\alpha \cdot B_0 + \beta \cdot S_0} \cdot y \, ,$$

i.e. the return of a portfolio equals the weighted average of the risk-free rate (the return of the zero-bond) and the return of the stock, where the weights correspond to the proportion invested in the zero-bond, i.e. $\frac{\alpha \cdot B_0}{\alpha \cdot B_0 + \beta \cdot S_0}$, and the proportion invested in the stock, i.e. $\frac{\beta \cdot S_0}{\alpha \cdot B_0 + \beta \cdot S_0}$, respectively.

2.1.17   Portfolios Generate Alternatives

Economically speaking, liquidating the portfolio $(\alpha, \beta) \in \mathbb{R}^2$ is equivalent to the outcome of having entered the financial contract described by the alternative $V_1^{(\alpha,\beta)} : \Omega \to \mathbb{R}$ leading to the payments $V_1^{(\alpha,\beta)}(g)$ and $V_1^{(\alpha,\beta)}(b)$ in the "good" and the "bad" states, respectively.

It will be crucial that even though the only claims which are traded are the bond and the stock, the formation of portfolios allows us to "generate" other alternatives which may be more desirable for us.

2.1.18   Attainability and Replication

An alternative $X : \Omega \to \mathbb{R}$ is said to be *attainable* if we can find a *replicating portfolio*, i.e. a portfolio (α, β) such that

$$V_1^{(\alpha,\beta)}(g) = X(g) \qquad \text{and} \qquad V_1^{(\alpha,\beta)}(b) = X(b)$$

both hold. A replicating portfolio is also called a *hedge portfolio*. One is said to have *hedged* a given alternative if one has set up a corresponding replicating portfolio. We could also call attainable alternatives *derivative financial instruments*, since they are potential financial contracts which may be derived from the basic traded securities, i.e. from the zero-bond and the stock.

2.1.19   Main Questions

In order to remain focused for the rest of the chapter we reformulate the questions we would like to answer for our model. Recall that our investor had defined a target alternative, say $X : \Omega \to \mathbb{R}$ which he wants to replicate by a suitable portfolio of traded securities, i.e. a portfolio containing positions in the zero-bond and the stock. We called such a portfolio a *replicating portfolio*. Now, the relevant questions.

a) Does every portfolio replicating a given alternative have the same initial cost?

b) Is the cost of replicating a non-zero claim, i.e. a positive alternative, always positive?

c) Can every alternative be replicated by a suitable portfolio?

d) In case that the value of a replicating portfolio is unique, can we find a simple way for determining the cost of replication?

The first question is the same as asking whether the *Law of One Price* holds. If the Law of One Price holds we are able to introduce the notion of the *fair value* of an alternative as the cost of replicating it.

The second question leads to the notion of an *arbitrage opportunity*, i.e. the opportunity to generate profits without risk and without any initial investment. Not too surprisingly, a market will not admit arbitrage opportunities if and only if the replication cost of non-zero claims is positive. As already noted, price systems for which non-zero claims have a positive fair value make economical sense since non-zero claims represent a real benefit with no drawbacks and there should thus be a cost to generating them.

The third question leads to the notion of *complete* markets, i.e. markets where every alternative is attainable. In our simple setting we prove that indeed every alternative can be replicated in a unique way by a suitable portfolio of traded securities. However, in more general models this need not be the case.

The last question will be answered by giving an explicit representation of the pricing functional. With such a representation one can determine the fair value of an alternative without explicit knowledge of a replicating portfolio. This formula is recast in the language of *equivalent martingale measures* in Section 2.4, where we show that it amounts to taking the expected value of the alternative discounted at the risk-free rate. This expected value will have to be taken with respect to a probability distribution Q on Ω which is generally different from the "natural" probability distribution P governing the randomness of our economy. The notion of arbitrage also turns out to be crucial in proving the existence of this new probability measure Q.

In our simple setting it is possible to address the first and third questions simultaneously.

2.2 Law of One Price, Completeness and Fair Value

If in an economy that is determined by some price system $(r, S_0, y(g), y(b))$, it is true that every alternative is attainable, we say that the market is *complete*.

2.2.1 Market Completeness

By definition the market is complete if and only if for each alternative $X : \Omega \to \mathbb{R}$ we find a portfolio $(\alpha, \beta) \in \mathbb{R}^2$ such that

$$\begin{bmatrix} V_1^{(\alpha,\beta)}(g) \\ V_1^{(\alpha,\beta)}(b) \end{bmatrix} = \begin{bmatrix} 100 & S_1(g) \\ 100 & S_1(b) \end{bmatrix} \begin{bmatrix} \alpha \\ \beta \end{bmatrix} = \begin{bmatrix} X(g) \\ X(b) \end{bmatrix} \tag{2.4}$$

Calculating the determinant of the matrix $\begin{bmatrix} 100 & S_1(g) \\ 100 & S_1(b) \end{bmatrix}$ yields

$$\begin{vmatrix} 100 & S_1(g) \\ 100 & S_1(b) \end{vmatrix} = 100 \cdot (S_1(b) - S_1(g)) \neq 0 .$$

Thus, the matrix is non-singular and (2.4) is uniquely solvable for each alternative $X \in \mathbb{R}^2$. Since the inverse of the above matrix is given by

$$\begin{bmatrix} \frac{-S_1(b)}{100 \cdot (S_1(g) - S_1(b))} & \frac{S_1(g)}{100 \cdot (S_1(g) - S_1(b))} \\ \frac{1}{S_1(g) - S_1(b)} & \frac{-1}{S_1(g) - S_1(b)} \end{bmatrix}$$

we find that the unique replicating portfolio (α_X, β_X) for X is given by

$$\begin{bmatrix} \alpha_X \\ \beta_X \end{bmatrix} = \begin{bmatrix} \frac{-S_1(b)}{100 \cdot (S_1(g) - S_1(b))} & \frac{S_1(g)}{100 \cdot (S_1(g) - S_1(b))} \\ \frac{1}{S_1(g) - S_1(b)} & \frac{-1}{S_1(g) - S_1(b)} \end{bmatrix} \begin{bmatrix} X(g) \\ X(b) \end{bmatrix}. \tag{2.5}$$

We have thus proved:

Theorem 2.1 *For any price system $\{r, S_0, y(g), y(b)\}$ the resulting market is complete.*

Replicating Arrow–Debreu securities

We now consider two simple claims, which are sometimes called *Arrow-Debreu claims* or *Arrow–Debreu securities*. They will be denoted by $E_1 : \Omega \to \mathbb{R}$ and $E_2 : \Omega \to \mathbb{R}$ and are defined by

$$E_1(g) \stackrel{\text{def}}{=} 1 \qquad \text{and} \qquad E_1(b) \stackrel{\text{def}}{=} 0,$$
$$E_2(g) \stackrel{\text{def}}{=} 0 \qquad \text{and} \qquad E_2(b) \stackrel{\text{def}}{=} 1.$$

The importance of Arrow–Debreu securities is that any alternative can be expressed in a unique way as a linear combination of Arrow–Debreu securities. Indeed, if X is an alternative, then we can write

$$X = X(g)E_1 + X(b)E_2. \tag{2.6}$$

On the other hand if X is given by $X = \lambda_g E_1 + \lambda_b E_2$, then by definition of E_1 and E_2 we immediately see that $X(g) = \lambda_g$ and $X(b) = \lambda_b$.
The following result can be easily verified by direct calculation.

Lemma 2.2 *The replicating portfolios for the Arrow–Debreu securities are given by*

$$(\alpha_{E_1}, \beta_{E_1}) = \left(\frac{-S_1(b)}{100 \cdot (S_1(g) - S_1(b))}, \frac{1}{S_1(g) - S_1(b)} \right), \tag{2.7}$$

and

$$(\alpha_{E_2}, \beta_{E_2}) = \left(\frac{S_1(g)}{100 \cdot (S_1(g) - S_1(b))}, \frac{-1}{S_1(g) - S_1(b)} \right), \tag{2.8}$$

respectively.

2.2.2 Replicating other Claims

Since (2.6) holds we can obtain the replicating portfolio (α_X, β_X) for an arbitrary alternative X by means of the equation

$$(\alpha_X, \beta_X) = X(g)(\alpha_{E_1}, \beta_{E_1}) + X(b)(\alpha_{E_2}, \beta_{E_2}) \,.$$

2.2.3 The Law of One Price and the Pricing Functional

Suppose we are given an alternative X. Then, there exists a unique portfolio (α_X, β_X) which replicates X, i.e. with

$$V_0^{(\alpha_X, \beta_X)} = X \,.$$

The value of this portfolio corresponds to the unique cost of generating the alternative X. It therefore makes sense to define the *fair value* or *fair price*, $\pi_0(X)$ of X as the cost of setting up this portfolio, i.e. by

$$\pi_0(X) \stackrel{\text{def}}{=} V_0^{(\alpha_X, \beta_X)} \,.$$

Obviously, the *pricing functional* $\pi_0 : \mathcal{A} \to \mathbb{R}$ is linear.

Valuing Arrow–Debreu securities

We now determine the value of π_0 on both Arrow–Debreu securities.

Lemma 2.3 *For the prices of the Arrow–Debreu securities the following holds.*

$$\pi_0(E_1) \;=\; \tfrac{1}{1+r} \cdot \tfrac{r-y(b)}{y(g)-y(b)} \;, \tag{2.9}$$

$$\pi_0(E_2) \;=\; \tfrac{1}{1+r} \cdot \tfrac{y(g)-r}{y(g)-y(b)} \,. \tag{2.10}$$

Proof The proof is straightforward. It uses the following facts:

a) the replicating portfolios for the Arrow–Debreu securities given by equations (2.7) and (2.8),

b) the definition of π_0, as well as

c) the expression $\frac{B_0}{100} = \frac{1}{1+r}$.

We illustrate this by carrying through the calculations for the case of E_1.

$$
\begin{aligned}
\pi_0(E_1) \;&=\; V_0^{\mathcal{R}(E_1)} = -\frac{S_1(b)}{100 \cdot (S_1(g) - S_1(b))} \cdot B_0 + \frac{1}{S_1(g) - S_1(b)} \cdot S_0 \\
&=\; -\frac{B_0}{100} \cdot \frac{S_0 \cdot (1 + y(b))}{S_0 \cdot (y(g) - y(b))} + \frac{1}{S_0 \cdot (y(g) - y(b))} \cdot S_0
\end{aligned}
$$

$$= -\frac{1}{1+r} \cdot \frac{1+y(b)}{y(g)-y(b)} + \frac{1}{y(g)-y(b)}$$

$$= \frac{1}{1+r} \cdot \left[\frac{1+r}{y(g)-y(b)} - \frac{1+y(b)}{y(g)-y(b)} \right]$$

$$= \frac{1}{1+r} \cdot \frac{r-y(b)}{y(g)-y(b)} .$$

□

An explicit formula for the pricing functional

As a linear functional, the pricing functional is completely determined by the values it takes on any basis of $\mathcal{A}$. As a corollary to (2.6) and Lemma 2.3 we obtain the following important result which describes the pricing functional in terms of the market parameters.

Theorem 2.4 *For any alternative* $X : \Omega \to \mathbb{R}$ *we have*

$$\pi_0(X) = X(g) \cdot \frac{1}{1+r} \cdot \frac{r-y(b)}{y(g)-y(b)} + X(b) \cdot \frac{1}{1+r} \cdot \frac{y(g)-r}{y(g)-y(b)} .$$

Proof For any alternative $X : \Omega \to \mathbb{R}$ we may write

$$X_1 = X(g) \cdot E_1 + X_1(b) \cdot E_2 .$$

Using the linearity of the pricing functional π_0 we obtain:

$$\pi_0(X) = X(g) \cdot \pi_0(E_1) + X(b) \cdot \pi_0(E_2) .$$

Now we obtain the desired result from the previous lemma.   □

Market consistency of the pricing functional

Note that if we take the fair value of our traded securities B_1 and S_1 we recover their market prices, i.e.

$$\pi_0(B_1) = B_0, \quad \text{and} \quad \pi_0(S_1) = S_0 .$$

This means that we are valuing alternatives in a way that is compatible with the market value of the basic securities!

2.3 Arbitrage and Positivity of the Pricing Functional

Note that we have not yet answered the question of whether the fair value of a claim — the cost of replicating a non-zero claim — is always positive. Economically, this would make sense since a claim represents a benefit and one would expect it to have a positive value. We can now offer a first answer in terms of our market parameters.

Proposition 2.5 *Let $(r, S_0, y(b), y(g))$ be a price system. Then, the fair value of every non-zero positive claim is strictly positive if and only if $y(b) < r < y(g)$ holds.*

Proof

Let $X : \Omega \to \mathbb{R}$ be an arbitrary claim. Then, by our representation formula in Theorem 2.4 we have:

$$\pi_0(X) = X(g) \cdot \frac{1}{1+r} \cdot \frac{r - y(b)}{y(g) - y(b)} + X(b) \cdot \frac{1}{1+r} \cdot \frac{y(g) - r}{y(g) - y(b)} .$$

Obviously, the right-hand side is strictly positive for every choice of non-zero positive X if and only if

$$\frac{1}{1+r} \cdot \frac{r - y(b)}{y(g) - y(b)} > 0 \qquad \text{and} \qquad \frac{1}{1+r} \cdot \frac{y(g) - r}{y(g) - y(b)} > 0$$

both hold. This is easily seen to be the case if and only if $y(b) < r < y(g)$ holds. This proves the assertion. $\qquad \square$

2.3.1 Arbitrage

Intuitively speaking an arbitrage opportunity enables any market participant to construct at no cost whatsoever a portfolio which can lead to potential gains and bears no risk of losses . Hence, our investor may virtually create wealth for himself starting from nothing. Arbitrage opportunities are sometimes called *free lunches*, since they offer a benefit for free. We now formalize this notion.

An *arbitrage opportunity* or *arbitrage portfolio* is any portfolio $(\alpha, \beta) \in \mathbb{R}^2$ satisfying the following two conditions:

a) $V_0^{(\alpha,\beta)} = 0$,

b) $V_1^{(\alpha,\beta)}(g) \geq 0$ and $V_1^{(\alpha,\beta)}(b) \geq 0$ with at least one of the inequalities being strict, i.e. ">".

Discounting by the stock

We may ask the question of whether or not we can find equivalent martingale measures with the stock as numeraire, i.e. a probability measure Q, such that for any alternative X, the value process $(\hat{\pi}(X)_{t=0,1})_{t=0,1}$ is a martingale. The proof of the following result is essentially the same as the proof of the existence of a risk-neutral probability.

Theorem 2.15 *There exist an equivalent martingale measure with respect to the stock if and only if the market admits no arbitrage opportunities. The martingale measure is unique and is denoted by $\hat{Q}$.*

From the above theorem we find that in absence of arbitrage there exists a measure $\hat{Q}$, such that

$$\pi(X) = S_0 \cdot E_{\hat{Q}}[\frac{X}{S_1}] .$$

2.6   Options and Forwards

In this final section we give a flavor of how the above results are applied to real life situations.

2.6.1   Call Options

A *call option* on the stock is a contract where the option buyer (the *holder*) has the right (but not the obligation) to buy at time $t = 1$ a pre-specified number of units of the stock at a pre-specified price of K currency units, called the *strike price*, from the option seller (the *writer*).

The payoff of a call

Assume the call entitles you to buy 1 unit of the stock. In the "good" state you will be able to buy the stock for K currency units, although its market value is $S_1(g)$. Of course, the holder will only exercise this right if $S_1(g) > K$. Thus, at time $t = 1$, in the "good" state, the economic effect of possessing the call is equivalent to receiving the *payoff* $X_{call}(g)$ given by

$$X_{call}(g) = \max\{0, S_1(g) - K\} .$$

Similarly, in the "bad" state, the economic effect of holding the call will be equivalent to receiving the payoff

$$X_{call}(b) = \max\{0, S_1(b) - K\} .$$

In other words: holding a call is economically equivalent to holding the claim X_{call}. Paying this claim to the holder of the option corresponds to a *cash-settlement* of the option. Should the market price of the stock at time $t = 1$ be above the strike price, the option writer will provide him with the cash equivalent of the price difference[6].

The fair value of a call

Applying equation 2.14 we immediately obtain:

$$\pi(X_{call}) = \frac{\max\{S_1(g) - K, 0\}}{1 + r} \cdot q + \frac{\max\{S_1(b) - K, 0\}}{1 + r} \cdot (1 - q)$$

with $q = \frac{r - y(b)}{y(g) - y(b)}$. This formula will be generalized to the multi-period case and goes back to the famous Cox–Ross–Rubinstein paper [17]. The multi-period version is widely used in the financial world to actually determine the price of options.

2.6.2   Put Options

A *put option* on the stock is a contract where the option buyer (the *holder*) has the right (but not the obligation) to sell at time $t = 1$ a pre-specified number of units of the stock at a pre-specified price of K currency units, called the *strike price*, to the option seller (the *writer*).

The payoff of a put

The payoff of the put is a contingent claim X_{put} which can be identified by similar arguments as for the call. If the put entitles the buyer to the purchase of 1 unit of the stock, the payoff of the put at maturity in the "good" state $X_{put}(g)$ will be given by

$$X_{put}(g) = \max\{0, K - S_1(g)\}$$

and in the "bad" state by

$$X_{put}(b) = \max\{0, K - S_1(b)\} \ .$$

The fair value of a put

As in the case of the call option, by applying equation 2.14, we find a simple formula for the fair value of a put option:

$$\pi(X_{put}) = \frac{\max\{K - S_1(g), 0\}}{1 + r} \cdot q + \frac{\max\{K - S_1(b), 0\}}{1 + r} \cdot (1 - q)$$

with $q = \frac{r - y(b)}{y(g) - y(b)}$.

---

[6]The exact contract specification could call for actual delivery of the stock by the option writer. However, cash-settlement is often more efficient. In any case, for valuation purposes it is economically equivalent.

2.6.3   Forward Contracts

A *forward contract* on the stock is an agreement to buy or sell the share at time $t = 1$ at a pre-specified price K, called the *delivery price*. The party who agrees to buy is said to have a *long position*. The other party is said to have a *short position*.

The payoff of a forward

The payoff of the forward contract at time $t = 1$ is represented by the alternative X_{for} given by

$$X_{for}(g) = K - S_1(g)$$

in the "good" state and by

$$X_{for}(b) = K - S_1(b)$$

in the "bad" state. In our terminology this is not quite a claim, because it may fail to be positive in all states of the world. However, we can represent it as a portfolio consisting of a bought put and a written call with strike K, i.e.

$$X_{for} = X_{put} - X_{call}. \tag{2.16}$$

The fair value of a forward

Again either by applying directly equation (2.14) or by using the fair value formulas for a call and a put derived above in conjunction with equation (2.16) we obtain:

$$\pi(X_{for}) = \frac{S_1(g) - K}{1 + r} \cdot q + \frac{S_1(b) - K}{1 + r} \cdot (1 - q)$$

with $q = \frac{r - y(b)}{y(g) - y(b)}$.

Concluding Remarks and Suggestions for Further Reading

In this chapter we have addressed within the simplest of settings many of the central topics of that part of mathematical finance dealing with arbitrage pricing of contingent claims . The rest of the book is essentially concerned with various generalizations of the arguments we have encountered here. The first generalization, in Chapter 6, will apply first to more general one-period models with an arbitrary finite number of securities. Then, in Chapters 9 and 11, we will deal with multi-period models with an arbitrary finite number of securities. The primary tool used in obtaining the main results for these more general models belongs to a special topic of linear algebra, the theory of positive linear functionals treated in

Chapter 3. The formulation of these more general models will be in the language of probability. We do not assume any prior knowledge of probability and develop all we need from the very beginning in Chapters 4, 5, 7, 8, 10, 13, and 15.

As already mentioned in the main body of this chapter, we do not deal with equilibrium pricing in this book. Treatments of this important part of mathematical finance can be found for instance in [20], [30], [41] or [47].

Chapter 3

Positive Linear Functionals

The proof that the absence of arbitrage implies the existence of a consistent positive linear pricing rule is more subtle and requires a separation theorem. The mathematical problem is equivalent to Farkas' Lemma of the alternative and to the basic duality theorem of linear programming.

Ph.H. Dybvig and S.A. Ross

In this chapter we develop some results of a slightly non-standard but important topic from linear algebra: positive linear functionals. This material constitutes the main technical tool for the proofs of the fundamental theorems of contingent claim pricing. Positive functionals play a prominent role in our considerations since they correspond to pricing functionals in arbitrage-free markets. The linear algebra needed here is summarized in Appendix A, where we also list some useful references.

3.1 Linear Functionals

Suppose $\mathbf{M}$ is a linear subspace of $\mathbb{R}^n$. A linear operator from $\mathbf{M}$ into $\mathbb{R}$ is usually called a *linear functional*.

From Proposition A.10 in Appendix A, we know that linear functionals are continuous functions. In particular, if $(\mathbf{x}_k)$ is a sequence in $\mathbf{M}$ converging to a vector $\mathbf{x} \in \mathbf{M}$ we have $\pi(\mathbf{x}_k) \to \pi(\mathbf{x})$ as k tends to infinity.

3.1.1   Linear Functionals on the Whole Space

Let $\pi : \mathbb{R}^n \to \mathbb{R}$ be a linear functional. The most common example of a linear functional is that of the linear functional induced by a vector $\mathbf{y} = (y_1, y_2, \ldots, y_n) \in \mathbb{R}^n$ by setting

$$\pi_{\mathbf{y}}(\mathbf{x}) \stackrel{\text{def}}{=} (\mathbf{x}|\mathbf{y}) = x_1 y_1 + x_2 y_2 + \ldots + x_n y_n$$

for any $\mathbf{x} = (x_1, x_2, \ldots, x_n) \in \mathbb{R}^n$. In fact, every linear functional on $\mathbb{R}^n$ can be represented in this way. This result is a so-called *representation theorem* for linear functionals.

Lemma 3.1 *Let $\pi : \mathbb{R}^n \to \mathbb{R}$ be a linear functional. Then there exists a unique vector $\mathbf{y}_\pi = (y_1, \ldots, y_n) \in \mathbb{R}$ such that*

$$\pi(\mathbf{x}) = (\mathbf{x}|\mathbf{y}_\pi) = x_1 y_1 + x_2 y_2 + \ldots + x_n y_n$$

holds for any $\mathbf{x} \in \mathbb{R}$.

Proof Set $\mathbf{y}_\pi \stackrel{\text{def}}{=} (\pi(\mathbf{e}_1), \ldots, \pi(\mathbf{e}_n))$. We then have

$$
\begin{aligned}
(\mathbf{x}|\mathbf{y}_\pi) &= x_1 y_1 + \ldots + x_n y_n \\
&= x_1 \pi(\mathbf{e}_1) + \ldots + x_n \pi(\mathbf{e}_n) \\
&= \pi(x_1 \mathbf{e}_1 + \ldots x_n \mathbf{e}_n) = \pi(\mathbf{x}),
\end{aligned}
$$

as claimed.   □

3.1.2   The Null Space of a Linear Functional

Consider a linear functional $\pi : \mathbf{M} \to \mathbb{R}$ defined on a (possibly proper) linear subspace $\mathbf{M}$ of $\mathbb{R}^n$. Recall that its *kernel* or *null space* is the linear subspace $N(\pi)$ consisting of all vectors that are mapped to 0 by π, i.e.

$$N(\pi) \stackrel{\text{def}}{=} \{\mathbf{x} \in \mathbf{M}; \pi(\mathbf{x}) = 0\} .$$

From the dimension formula (Proposition A.16) we know that if the subspace $\mathbf{M}$ is k-dimensional, then the kernel of π is a $(k-1)$-dimensional linear subspace of $\mathbf{M}$ provided $\pi \neq 0$. The following result will prove useful. It tells us that the kernel uniquely determines a linear functional up to a scalar factor, which is the key to constructing extensions of π to the whole space, should $\mathbf{M}$ be a proper subspace of $\mathbb{R}^n$.

Lemma 3.2 *Let π and ψ be linear functionals defined on a linear subspace $\mathbf{M}$ of $\mathbb{R}^n$ with $\mathbf{Z} \stackrel{\text{def}}{=} N(\pi) = N(\psi)$. Then, there exists a real number λ such that $\pi = \lambda \psi$.*

Proof Without loss of generality we can assume that π and ψ are not equal to the zero functional. Let k be the dimension of $\mathbf{M}$. Then $\mathbf{Z}$ is $(k-1)$-dimensional. By the discussion in A.1.5 we can thus write $\mathbf{M}$ as the direct sum of $\mathbf{Z}$ and a one-dimensional subspace $\mathbf{U} \stackrel{\text{def}}{=} span(\mathbf{x}_0)$ spanned by a vector $\mathbf{x}_0 \in \mathbf{M} \setminus \mathbf{Z}$, i.e.

$$\mathbf{M} = \mathbf{Z} \oplus \mathbf{U} .$$

Since $\mathbf{x}_0 \notin \mathbf{Z}$ we can find a $\lambda \neq 0$ such that $\pi(\mathbf{x}_0) = \lambda\psi(\mathbf{x}_0)$. Writing any vector $\mathbf{x} \in \mathbf{M}$ as $\mathbf{z} + \alpha\mathbf{x}_0$ we get

$$
\begin{aligned}
\pi(\mathbf{x}) \;\; &= \pi(\mathbf{z} + \alpha\mathbf{x}_0) = \pi(\mathbf{z}) + \alpha\pi(\mathbf{x}_0) = \alpha\pi(\mathbf{x}_0) \\
&= \alpha\lambda\psi(\mathbf{x}_0) = \lambda\psi(\mathbf{z}) + \lambda\psi(\alpha\mathbf{x}_0) = \lambda\psi(\mathbf{z} + \alpha\mathbf{x}_0) \\
&= \lambda\psi(\mathbf{x}) .
\end{aligned}
$$

$\square$

3.1.3   Extensions of Linear Functionals

Let $\mathbf{M}$ be a k-dimensional linear subspace of $\mathbb{R}^n$ and $\pi : \mathbf{M} \to \mathbb{R}$ a given linear functional. A linear functional $\tilde{\pi} : \mathbb{R}^n \to \mathbb{R}$ is said to be an *extension* of π if

$$\tilde{\pi}(\mathbf{x}) = \pi(\mathbf{x})$$

holds for all $\mathbf{x} \in \mathbf{M}$. The following result characterizes extensions of a given functional.

Lemma 3.3 *If $\tilde{\pi} : \mathbb{R}^n \to \mathbb{R}$ is an extension of π and $\tilde{\mathbf{z}}$ is its representation vector as in* Lemma 3.1, *then $\tilde{\mathbf{z}}$ belongs to the orthogonal complement of $N(\pi)$, i.e. $(\mathbf{x}|\tilde{\mathbf{z}}) = 0$ for all $\mathbf{x} \in N(\pi)$. Conversely, let $\tilde{\mathbf{z}}$ be any vector orthogonal to $N(\pi)$ and define $\tilde{\pi} : \mathbb{R}^n \to \mathbb{R}$ by $\tilde{\pi}(\mathbf{x}) \stackrel{\text{def}}{=} (\mathbf{x}|\tilde{\mathbf{z}})$. Then there exists a unique $\lambda \in \mathbb{R}$ such that $\lambda\tilde{\pi}$ is an extension of π.*

Proof If $\tilde{\pi}$ is an extension we get in particular $(\mathbf{x}|\tilde{\mathbf{z}}) = \tilde{\pi}(\mathbf{x}) = \pi(\mathbf{x}) = 0$ for all $\mathbf{x} \in N(\pi)$, proving the first statement. For the second statement let $\tilde{\mathbf{z}}$ be orthogonal to $N(\pi)$. Then, $\tilde{\pi}_\mathbf{M} : \mathbf{M} \to \mathbb{R}$, the restriction of $\tilde{\pi}$ to $\mathbf{M}$, has the same kernel as π. The claim now follows from Lemma 3.2. $\square$

We have thus identified the set of extensions of π as being in one-to-one correspondence with the directions orthogonal to $N(\pi)$.

3.2   Positive Linear Functionals Introduced

Before introducing the notion of a positive linear functional we need to develop the concept of positivity for vectors in Euclidean space.

3.2.1   Natural Order on $\mathbb{R}^n$

Euclidean space has a natural *order structure*. It is induced by the following notion of positivity.

- $\mathbf{x} \in \mathbb{R}^n$ is said to be *positive*, or *non-negative*, if $x_i \geq 0$ holds for all $1 \leq i \leq n$. In this case we write $\mathbf{x} \geq 0$.

- $\mathbf{x} \in \mathbb{R}^n$ is said to be *strictly positive* if it is positive but non-zero. In this case we write $\mathbf{x} > 0$.

- $\mathbf{x} \in \mathbb{R}^n$ is said to be *strongly positive* if $\mathbf{x}_i > 0$ holds for all $1 \leq i \leq n$. In this case we write $\mathbf{x} \gg 0$.

The order structure implied by these positivity concepts is then summarized by the following definitions. For $\mathbf{x}, \mathbf{y} \in \mathbb{R}^n$ we write

- $\mathbf{x} \leq \mathbf{y}$ whenever $0 \leq \mathbf{y} - \mathbf{x}$;

- $\mathbf{x} < \mathbf{y}$ whenever $0 < \mathbf{y} - \mathbf{x}$; and

- $\mathbf{x} \ll \mathbf{y}$ whenever $0 \ll \mathbf{y} - \mathbf{x}$.

Remark 3.4 *If $n = 1$, i.e. in the case of the real line $\mathbb{R}$, the notions of strict and strong positivity coincide.*

3.2.2   Bases of Positive Vectors

The standard basis of $\mathbb{R}^n$, denoted by $\mathbf{e}_1, \dots \mathbf{e}_n$, consists of strictly (but not strongly!) positive vectors. We will show that as long as a subspace contains one strongly positive vector (which is certainly the case for $\mathbb{R}^n$) we can also find a basis consisting of strongly positive vectors. This result will be key when we construct positive extensions of positive linear functionals defined on a proper subspace of $\mathbb{R}^n$.

First we prove that the set of strongly positive vectors of a subspace is always open in the topology of that subspace. Hence, near enough to a strongly positive vector all other vectors are also strongly positive.

Lemma 3.5 *Let $\mathbf{Z}$ be a linear subspace of $\mathbb{R}^n$ containing a strongly positive vector $\mathbf{x}$. Then, there is an $\epsilon > 0$ such that all vectors in the set $\{\mathbf{z} \in \mathbf{Z}; |\mathbf{z} - \mathbf{x}| < \epsilon\}$ are also strongly positive, i.e. the set $\{\mathbf{z} \in \mathbf{Z}; \mathbf{z} \gg 0\}$ of strongly positive vectors in $\mathbf{Z}$ is open in $\mathbf{Z}$.*

Proof Take $\epsilon > 0$ such that $x_i > \epsilon$ for all $i = 1, \dots, n$. For any $\mathbf{z} \in \mathbb{R}^n$ satisfying $|\mathbf{z} - \mathbf{x}| < \epsilon$ we have

$$\epsilon > |\mathbf{z} - \mathbf{x}| \geq |z_i - x_i| \geq x_i - z_i$$

for all $i = 1, \ldots, n$. It follows that $z_i > x_i - \epsilon > 0$ and therefore that $\mathbf{z} \gg 0$. This means that $\{\mathbf{z} \in \mathbb{R}^n; |\mathbf{z} - \mathbf{x}| < \epsilon\}$ is open. Since

$$\{\mathbf{z} \in \mathbf{Z}; |\mathbf{z} - \mathbf{x}| < \epsilon\} = \{\mathbf{z} \in \mathbb{R}^n; |\mathbf{z} - \mathbf{x}| < \epsilon\} \cap \mathbf{Z}$$

the claim follows. $\qquad\square$

Theorem 3.6 *Let $\mathbf{Z}$ be a k-dimensional linear subspace of $\mathbb{R}^n$. If there exists a strongly positive vector $\mathbf{x} \in \mathbf{Z}$, then there exists a basis $\{\mathbf{z}_1, \mathbf{z}_2, \ldots, \mathbf{z}_k\}$ of $\mathbf{Z}$ consisting solely of strongly positive vectors.*

Proof Let $\mathbf{z}_1 \in \mathbf{Z}$ be strongly positive. Complete $\{\mathbf{z}_1\}$ to a basis

$$\{\mathbf{z}_1, \tilde{\mathbf{z}}_2, \ldots, \tilde{\mathbf{z}}_k\}$$

and set

$$\mathbf{z}_j(\lambda) \overset{\text{def}}{=} \lambda \mathbf{z}_1 + (1 - \lambda)\tilde{\mathbf{z}}_j$$

for $j = 2, \ldots, k$. We claim that for any $\lambda \in [0, 1)$ the set $\{\mathbf{z}_1, \mathbf{z}_2(\lambda), \ldots, \mathbf{z}_k(\lambda)\}$ is also a basis. Indeed, it suffices to show its linear independence. Assume that

$$\alpha_1 \mathbf{z}_1 + \alpha_2 \mathbf{z}_2(\lambda) + \ldots + \alpha_k \mathbf{z}_k(\lambda) = 0$$

holds. Rewriting this yields

$$[\alpha_1 + \lambda\alpha_2 + \ldots + \lambda\alpha_k]\mathbf{z}_1 + \alpha_2(1 - \lambda)\tilde{\mathbf{z}}_2 + \ldots + \alpha_k(1 - \lambda)\tilde{\mathbf{z}}_k = 0 .$$

Because $\{\mathbf{z}_1, \tilde{\mathbf{z}}_2, \ldots, \tilde{\mathbf{z}}_k\}$ is a basis it follows that

$$\alpha_1 + \lambda\alpha_2 + \ldots + \lambda\alpha_k = \alpha_2(1 - \lambda) = \ldots = \alpha_k(1 - \lambda) = 0 .$$

From $\lambda \neq 1$ we obtain

$$\alpha_1 = \alpha_2 = \ldots = \alpha_k = 0 ,$$

proving that $\{\mathbf{z}_1, \mathbf{z}_2(\lambda), \ldots, \mathbf{z}_k(\lambda)\}$ is a basis for $\mathbf{Z}$. Observe now that

$$|\mathbf{z}_1 - \mathbf{z}_j(\lambda)| = |\mathbf{z}_1 - (\lambda\mathbf{z}_1 + (1 - \lambda)\tilde{\mathbf{z}}_j)| = (1 - \lambda)|\mathbf{z}_1 - \tilde{\mathbf{z}}_j|$$

holds for $j = 2, \ldots, k$. Choosing λ close enough to 1 we may get all of the vectors $\mathbf{z}_2(\lambda), \ldots, \mathbf{z}_k(\lambda)$ as near to $\mathbf{z}_1$ as we wish. Since $\mathbf{z}_1$ is strongly positive, by Lemma 3.5 these vectors will also be strongly positive if λ is close enough to 1. $\qquad\square$

The idea of the above lemma is, of course, that the set $\{\mathbf{x} \in \mathbf{Z}; \mathbf{x} \gg 0\}$ is open in $\mathbf{Z}$. Thus, if it is nonempty there will be enough "directions" in it to choose a basis from.

3.3   Separation Theorems

The results of this section all have a fairly intuitive geometric interpretation. We will try to emphasize it throughout.

3.3.1   Separating Hyperplanes

Given $\pi : \mathbb{R}^n \to \mathbb{R}$, a non-zero linear functional, let $\mathbf{y}_\pi \in \mathbb{R}^n$ be its representation vector, i.e. $\pi(\mathbf{x}) = (\mathbf{x}|\mathbf{y}_\pi)$ for all $\mathbf{x} \in \mathbb{R}^n$.

For any $b \in \mathbb{R}$ we define

- $H_\pi(b) \overset{\text{def}}{=} \{\mathbf{x} \in \mathbb{R}; \pi(\mathbf{x}) = (\mathbf{x}|\mathbf{y}_\pi) = b\}$,

- $H_\pi^+(b) \overset{\text{def}}{=} \{\mathbf{x} \in \mathbb{R}^n; \pi(\mathbf{x}) = (\mathbf{x}|\mathbf{y}_\pi) \geq b\}$, and

- $H_\pi^-(b) \overset{\text{def}}{=} \{\mathbf{x} \in \mathbb{R}^n; \pi(\mathbf{x}) = (\mathbf{x}|\mathbf{y}_\pi) \leq b\}$.

Note that $H_\pi(0)$ corresponds to the null space $N(\pi)$. We can picture $H_\pi(b)$ as *separating* Euclidean space $\mathbb{R}^n$ into two parts $H_\pi^+(b)$ and $H_\pi^-(b)$, called the positive and negative *half-spaces* associated with (π, b), respectively. The positive and negative half-spaces have only the set $H_\pi(b)$ in common:

$$H_\pi(b) = H_\pi^+(b) \cap H_\pi^-(b). \tag{3.1}$$

If $\mathbf{A}$ and $\mathbf{B}$ are subsets of $\mathbb{R}^n$ we write

$$\mathbf{A} + \mathbf{B} \overset{\text{def}}{=} \{\mathbf{x} + \mathbf{y}; \mathbf{x} \in \mathbf{A} \text{ and } \mathbf{y} \in \mathbf{B}\} \ .$$

We call $\mathbf{A} + \mathbf{B}$ the *translation* of $\mathbf{A}$ by $\mathbf{B}$. If $\mathbf{A} = \{\mathbf{a}\}$ for some vector $\mathbf{a}$, we just write $\mathbf{a} + \mathbf{B}$.

Lemma 3.11 *Let $\mathbf{x}_b$ be any vector satisfying $\pi(\mathbf{x}_b) = b$. Then*

a) $H_\pi(b) = \mathbf{x}_b + H_\pi(0)$;

b) $H_\pi^+(b) = \mathbf{x}_b + H_\pi^+(0)$; and

c) $H_\pi^-(b) = \mathbf{x}_b + H_\pi^-(0)$.

Proof Take $\mathbf{x}_b$ to be any vector satisfying $\pi(\mathbf{x}_b) = b$. Then

- $\pi(\mathbf{y}) \geq 0$ implies that $\pi(\mathbf{x}_b + \mathbf{y}) = b + \pi(\mathbf{y}) \geq b$, and

- if $\pi(\mathbf{z}) \geq b$ and we set $\mathbf{y} \overset{\text{def}}{=} \mathbf{z} - \mathbf{x}_b$, we have $\mathbf{z} = \mathbf{x}_b + \mathbf{y}$ and $\pi(\mathbf{y}) = \pi(\mathbf{z}) - \pi(\mathbf{x}_b) = \pi(\mathbf{z}) - b \geq 0$.

These two observations prove that $H_\pi^+(b) = \mathbf{x}_b + H_\pi^+(0)$. The proofs for $H_\pi^-(b)$ and $H_\pi(b)$ are very much the same. □

Recall that the null space of π, $H_\pi(0)$, is a linear subspace of dimension $n-1$, i.e. a *linear hyperplane*. The above result tells us that $H_\pi(b)$ is an *affine hyperplane*, i.e. a linear hyperplane translated by a fixed vector. We shall use the term *hyperplane* to mean affine hyperplane. Note that a linear hyperplane is also an affine hyperplane. Hence, it is customary to say that the hyperplane $H_\pi(b)$ separates $\mathbb{R}^n$ into the two half-spaces $H_\pi^+(b)$ and $H_\pi^-(b)$.

Recalling the link between inner products and angles (see Appendix A) we see that vectors belonging to $H_\pi^+(0)$ form an angle greater than 0 and smaller than 180 with $\mathbf{y}_\pi$. On the other hand, vectors in $H_\pi^-(0)$ form an angle greater than 180 and smaller than 360 with $\mathbf{y}_\pi$.

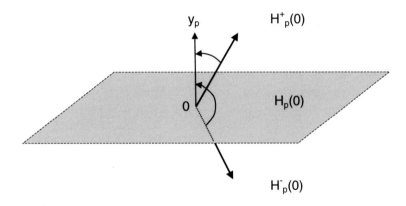

Figure 3.1: Separating hyperplane.

We say that two subsets $\mathbf{A}$ and $\mathbf{B}$ are *separated* by the hyperplane $H_\pi(b)$ if $\mathbf{A}$ lies on one side of $H_\pi(b)$ and $\mathbf{B}$ on the other, i.e. either

- $\pi(\mathbf{x}) \geq b \geq \pi(\mathbf{y})$; or

- $\pi(\mathbf{x}) \leq b \leq \pi(\mathbf{y})$

holds for all $\mathbf{x} \in \mathbf{A}$ and $\mathbf{y} \in \mathbf{B}$.

If additionally neither $\mathbf{A}$ nor $\mathbf{B}$ intersect the hyperspace $H_\pi(0)$ — i.e. if either

- $\pi(\mathbf{x}) > b > \pi(\mathbf{y})$; or

- $\pi(\mathbf{x}) < b < \pi(\mathbf{y})$

holds for all $\mathbf{x} \in \mathbf{A}$ and $\mathbf{y} \in \mathbf{B}$ — we say that $\mathbf{A}$ and $\mathbf{B}$ are *strictly separated* by $H_\pi(b)$.

Sometimes we shall say that the linear functional π *(strictly) separates* the sets $\mathbf{A}$ and $\mathbf{B}$ if there exists a $b \in \mathbb{R}$ such that $\mathbf{A}$ and $\mathbf{B}$ are (strictly) separated by $H_\pi(b)$. We will prove that $H_\pi(b)$, $H_\pi^+(b)$ and $H_\pi^-(b)$ are all convex, closed subsets of $\mathbb{R}^n$. This will be the consequence of a simple lemma about translations of convex and closed sets.

Proposition 3.12 *Let $\mathbf{A}$ and $\mathbf{B}$ be subsets of $\mathbb{R}^n$. Then:*

- *If both $\mathbf{A}$ and $\mathbf{B}$ are convex, then the same is true for $\mathbf{A} + \mathbf{B}$.*

- *If $\mathbf{A}$ is compact and $\mathbf{B}$ is closed, then $\mathbf{A} + \mathbf{B}$ is closed.*

Proof (i) For $j = 1, 2$ let $\mathbf{y}_j = \mathbf{x}_j + \mathbf{z}_j \in \mathbf{A} + \mathbf{B}$ and $\lambda \in [0, 1]$. Then by convexity of $\mathbf{A}$ and $\mathbf{B}$ we get $\lambda \mathbf{x}_1 + (1 - \lambda)\mathbf{x}_2 \in \mathbf{A}$ and $\lambda \mathbf{z}_1 + (1 - \lambda)\mathbf{z}_2 \in \mathbf{B}$. Hence, $\lambda \mathbf{y}_1 + (1 - \lambda)\mathbf{y}_2 = \lambda \mathbf{x}_1 + (1 - \lambda)\mathbf{x}_2 + \lambda \mathbf{z}_1 + (1 - \lambda)\mathbf{z}_2 \in \mathbf{A} + \mathbf{B}$. This proves convexity of $\mathbf{A} + \mathbf{B}$.

(ii) Let $(\mathbf{x}_j + \mathbf{z}_j)$ be a sequence in $\mathbf{A} + \mathbf{B}$ converging to some $\mathbf{y} \in \mathbb{R}^n$. Since $\mathbf{A}$ is compact, we may select a subsequence $(\mathbf{x}_{j_k})$ such that $\lim_{k \to \infty} \mathbf{x}_{j_k} = \mathbf{x}$ for some $\mathbf{x} \in \mathbf{A}$. But then $\lim_{k \to \infty} \mathbf{z}_{j_k} = \mathbf{y} - \mathbf{x}$. Since $\mathbf{B}$ is closed it follows that $\mathbf{z} \overset{\text{def}}{=} \mathbf{y} - \mathbf{x} \in \mathbf{B}$. Hence, $\mathbf{y} = \mathbf{x} + (\mathbf{y} - \mathbf{x}) = \mathbf{x} + \mathbf{z} \in \mathbf{A} + \mathbf{B}$ proving that $\mathbf{A} + \mathbf{B}$ is closed.   □

Lemma 3.13 *$H_\pi(b)$, $H_\pi^+(b)$ and $H_\pi^-(b)$ are all convex, closed subsets of $\mathbb{R}^n$ and*

$$H_\pi^+(b) \cap H_\pi^-(b) = H_\pi(b) .$$

Proof That the intersection of $H_\pi^+(b)$ and $H_\pi^-(b)$ equals $H_\pi(b)$ was already mentioned in (3.1).

To verify convexity of $H_\pi^+(b)$ take $\mathbf{x}, \mathbf{y} \in H_\pi^+(b)$ and $0 \leq \lambda \leq 1$ and just note that

$$\pi(\lambda \mathbf{x} + (1 - \lambda)\mathbf{y}) = \lambda \pi(\mathbf{x}) + (1 - \lambda)\pi(\mathbf{y}) \geq \lambda b + (1 - \lambda)b = b .$$

To verify that $H_\pi^+(b)$ is closed take a sequence $(\mathbf{x}_j)$ in $H_\pi^+(b)$ converging to an $\mathbf{x} \in \mathbb{R}^n$. Since by continuity of π we have that $\pi(\mathbf{x}_j) \to \pi(\mathbf{x})$ as $j \to \infty$ and by assumption $\pi(\mathbf{x}_j) \geq b$ for all j, we obtain that $\pi(\mathbf{x}) \geq b$, and, hence, that $H_\pi^+(b)$ is closed.

The proof for $H_\pi^-(b)$ uses opposite inequality signs. The claim for $H_\pi(b)$ can be either obtained in a similar manner or just by noting that by (3.1) it is the intersection of two closed convex sets, which immediately yields the desired result.   □

3.3.2 Separation Theorems

Separation theorems give conditions under which we can separate two given subsets of $\mathbb{R}^n$ by a linear functional. Recall the well-known result in analysis stating that, on a compact set, a continuous function achieves its minimum and its maximum. This fact will be used next to prove that we can always strictly separate the set $\{0\}$ from a closed convex subset which does not contain 0.

Theorem 3.14 *Let* $\mathbf{C} \subset \mathbb{R}^n$ *be a closed convex set which does not contain the origin, i.e.* $0 \notin \mathbf{C}$. *Then there exists a vector* $\mathbf{x}_0 \in \mathbf{C}$ *such that*

$$(\mathbf{x}_0|\mathbf{x}) \geq |\mathbf{x}_0|^2$$

holds for each $\mathbf{x} \in \mathbf{C}$.

Proof Let $r > 0$ be such that $\mathbf{C}_r \stackrel{\text{def}}{=} \{\mathbf{x} \in \mathbf{C}\,;\, |\mathbf{x}| \leq r\} \neq \emptyset$. Then $\mathbf{C}_r$ is compact and convex. Since, $\mathbf{x} \mapsto |\mathbf{x}|$ is continuous this function achieves its minimum on $\mathbf{C}_r$, i.e. we find $\mathbf{x}_0 \in \mathbf{C}$ such that

$$|\mathbf{x}_0| \leq |\mathbf{x}|$$

holds for all $\mathbf{x} \in \mathbf{C}_r$ and therefore for all $\mathbf{x} \in \mathbf{C}$. By the convexity of $\mathbf{C}$ we have

$$\lambda\mathbf{x} + (1-\lambda)\mathbf{x}_0 = \mathbf{x}_0 + \lambda(\mathbf{x} - \mathbf{x}_0) \in \mathbf{C}$$

for $\lambda \in [0, 1]$ and $\mathbf{x} \in \mathbf{C}$. Hence,

$$|\mathbf{x}_0|^2 \leq |\mathbf{x}_0 + \lambda(\mathbf{x} - \mathbf{x}_0)|^2 = |\mathbf{x}_0|^2 + 2\lambda(\mathbf{x}_0|\mathbf{x} - \mathbf{x}_0) + \lambda^2|\mathbf{x} - \mathbf{x}_0|^2\,.$$

It follows that

$$0 \leq 2\lambda(\mathbf{x}_0|\mathbf{x} - \mathbf{x}_0) + \lambda^2|\mathbf{x} - \mathbf{x}_0|^2$$

and, hence,

$$0 \leq 2(\mathbf{x}_0|\mathbf{x} - \mathbf{x}_0) + \lambda|\mathbf{x} - \mathbf{x}_0|^2\,.$$

Letting $\lambda \to 0$ we get

$$0 \leq (\mathbf{x}_0|\mathbf{x} - \mathbf{x}_0) = (\mathbf{x}|\mathbf{x}_0) - |\mathbf{x}_0|^2\,.$$

We thus obtain: $|\mathbf{x}_0|^2 \leq (\mathbf{x}|\mathbf{x}_0)$ for any $\mathbf{x} \in \mathbf{C}$. $\qquad\square$

As a corollary we state the separation of 0 and the convex set $\mathbf{C}$ explicitly.

Corollary 3.15 *Let* $\mathbf{C} \subset \mathbb{R}^n$ *be a closed convex set which does not contain the origin, i.e.* $0 \notin \mathbf{C}$. *Then there exists a linear functional* $\pi : \mathbb{R}^n \to \mathbb{R}$ *and a real number* b *such that* $H_\pi(b)$ *strictly separates* 0 *from* $\mathbf{C}$.

A geometric version of Farkas' lemma

Before giving the standard form of Farkas' Lemma we shall prove a geometric version of it.

Theorem 3.21 *Let* $\mathbf{K} \subset \mathbb{R}^m$ *be a closed and convex cone. Then* $\mathbf{b} \in \mathbb{R}^m$ *belongs to* $\mathbf{K}$ *if and only if for each* $\mathbf{y} \in \mathbf{K}$ *satisfying* $(\mathbf{x}|\mathbf{y}) \geq 0$ *for all* $\mathbf{x} \in \mathbf{K}$ *we also have* $(\mathbf{b}|\mathbf{y}) \geq 0$.

Proof
The "if" part is obvious. To prove the opposite direction assume that $\mathbf{b}$ does not belong to $\mathbf{K}$. Set

$$\mathbf{K_b} \overset{\text{def}}{=} \{\mathbf{x} - \mathbf{b}; \mathbf{x} \in \mathbf{K}\} .$$

Thus, $\mathbf{K_b}$ is the translation of $\mathbf{K}$ by $\{-\mathbf{b}\}$. By Lemma 3.12 we see that $\mathbf{K_b}$ is closed and convex. Moreover, since $\mathbf{b}$ does not belong to $\mathbf{K}$, also 0 does not belong to $\mathbf{K_b}$.
By the Theorem 3.14 there exists a $\mathbf{y}_0 \in \mathbf{K}_b$ such that

$$(\mathbf{z}|\mathbf{y}_0) \geq |\mathbf{y}_0|^2 > 0$$

holds for all $\mathbf{z} \in \mathbf{K_b}$, which is the same as saying that

$$(\mathbf{x} - \mathbf{b}|\mathbf{y}_0) \geq |\mathbf{y}_0|^2 > 0$$

holds for all $\mathbf{x} \in \mathbf{K}$. Hence,

$$(\mathbf{x}|\mathbf{y}_0) > (\mathbf{b}|\mathbf{y}_0)$$

holds for all $\mathbf{x} \in \mathbf{K}$. In particular choosing $\mathbf{x} = 0$ we see that

$$(\mathbf{b}|\mathbf{y}_0) < 0$$

holds. On the other hand, since for each $\mathbf{x} \in \mathbf{K}$ and $\lambda > 0$, also $\lambda \cdot \mathbf{x} \in \mathbf{K}$, we see that

$$\lambda(\mathbf{x}|\mathbf{y}_0) > (\mathbf{b}|\mathbf{y}_0)$$

and therefore

$$(\mathbf{x}|\mathbf{y}_0) > \frac{(\mathbf{b}|\mathbf{y}_0)}{\lambda}$$

holds for all $\mathbf{x} \in \mathbf{K}$ and $\lambda > 0$. Letting λ tend to infinity we obtain

$$(\mathbf{x}|\mathbf{y}_0) \geq 0$$

for all $\mathbf{x} \in \mathbf{K}$. It follows that if no positive solution exists, there then exists a $\mathbf{y}_0 \in K_\mathbf{b}$ such that $(\mathbf{x}|\mathbf{y}_0) \geq 0$ for all $\mathbf{x} \in \mathbf{K}$ but $(\mathbf{b}|\mathbf{y}_0) < 0$. This proves the theorem.   □

The geometric interpretation is the following. Assume that $\mathbf{y} \in \mathbb{R}^m$ is given and denote by π_y the associated linear functional, i.e. $\pi_{\mathbf{y}}(\mathbf{x}) \overset{\text{def}}{=} (\mathbf{x}|\mathbf{y})$. Recall that

$(\mathbf{x}|\mathbf{y}) \geq 0$ for some $\mathbf{x}$ means that $\mathbf{x}$ is contained in the positive half-space $H^+_{\pi_{\mathbf{y}}}(0)$. Hence, the above theorem says that $\mathbf{b}$ belongs to $\mathbf{K}$ if and only if $\mathbf{b}$ is contained in every half-space which contains $\mathbf{K}$. We can thus formulate it as follows.

Corollary 3.22 *A closed and convex cone is always equal to the intersection of all half-spaces containing it.*

Farkas' Lemma

The following is a more traditional version of Farkas' lemma.

Theorem 3.23 *The linear system (3.2) has a positive solution if and only if the following holds: for any $\mathbf{y} \in \mathbb{R}^m$ satisfying*

$$(A\mathbf{x}|\mathbf{y}) \geq 0$$

for all $\mathbf{x} \in \mathbb{R}^m_+$, we also have

$$(\mathbf{b}|\mathbf{y}) \geq 0 \, .$$

Proof The linear system (3.2) admits a positive solution if and only if $\mathbf{b}$ belongs to $\mathbf{K}_A = \{A\mathbf{x}; \mathbf{x} \in \mathbb{R}^n_+\}$ which by Corollary 3.20 is a closed and convex cone. By Theorem 3.21 this is the case if and only if for each $\mathbf{y}$ satisfying $(\mathbf{z}|\mathbf{y}) \geq 0$ for all $\mathbf{z} \in \mathbf{K}_A$ we also have $(\mathbf{b}|\mathbf{y}) \geq 0$. This is equivalent to saying that for each $\mathbf{y}$ satisfying $(A\mathbf{x}|\mathbf{y}) \geq 0$ for all $\mathbf{x} \in \mathbb{R}^n_+$ we also have $(\mathbf{b}|\mathbf{y}) \geq 0$. $\qquad \square$

3.4 Extension of Positive Linear Functionals

Let $\mathbf{M}$ be a k-dimensional linear subspace of $\mathbb{R}^n$ and suppose that $\pi : \mathbf{M} \to \mathbb{R}$ is a positive linear functional. Can we find extensions of π which are also positive? The next lemma characterizes positive extensions.

Lemma 3.24 *There exists a (strongly) positive extension of the (strongly) positive non-zero linear functional $\pi : \mathbf{M} \to \mathbb{R}$ if and only if $N(\pi)^\perp$, the orthogonal complement of its null space $N(\pi)$, contains a (strongly) positive non-zero vector.*

Proof Let $\tilde{\pi}$ be a (strongly) positive extension of π. Then by Lemma 3.9 its representation vector $\mathbf{y}_{\tilde{\pi}}$ is (strongly) positive and non-zero. Moreover, for any $\mathbf{x} \in N(\pi)$ we have $(\mathbf{x}|\mathbf{y}_{\tilde{\pi}}) = \tilde{\pi}(\mathbf{x}) = \pi(\mathbf{x}) = 0$. Therefore, $\mathbf{y}_{\tilde{\pi}}$ is a (strongly) positive non-zero vector contained in $N(\pi)^\perp$.

On the other hand let $\mathbf{y}$ be a (strongly) positive non-zero vector in $N(\pi)^\perp$. Again, by Lemma 3.9, we know that there exists a non-zero number λ such that the linear functional $\pi_{\lambda\mathbf{y}}$ defined by $\pi_{\lambda\mathbf{y}}(\mathbf{x}) \overset{\text{def}}{=} \lambda(\mathbf{x}|\mathbf{y})$ is an extension of π. Because of the (strong) positivity of π we conclude that $\lambda > 0$. It follows that $\pi_{\mathbf{y}}$ is a (strongly) positive extension of π. $\qquad \square$

We shall now narrow down our focus and concentrate only on the existence of strongly positive extensions of strongly positive linear functionals. Hence, we assume that π is strongly positive. By the above lemma π does have a strongly positive extension if and only if we can find a strongly positive vector in the orthogonal complement of its null space $N(\pi)$. That this is always the case will be a consequence of the next result.

Proposition 3.25 *Let $\mathbf{Z}$ be a proper linear subspace of $\mathbb{R}^n$ such that $\mathbf{Z} \cap \mathbb{R}^n_+ = \{0\}$. Then, $\mathbf{Z}^\perp$ has a basis consisting of strongly positive vectors.*

Proof Define $\mathbf{C}$ to be the smallest convex set containing the standard basis vectors $\mathbf{e}_1, \ldots, \mathbf{e}_k$. Then, $\mathbf{C}$ is convex, compact and consists entirely of strictly positive elements. In particular, $0 \notin \mathbf{C}$. By assumption $\mathbf{C} \cap \mathbf{Z} = \emptyset$. Applying Theorem 3.16 we obtain the existence of an $\mathbf{x}_0 \in \mathbf{C}$ such that

$$(\mathbf{x}_0|\mathbf{x}) \geq |\mathbf{x}_0|^2$$

for each $\mathbf{x} \in \mathbf{C}$ and

$$(\mathbf{x}_0|\mathbf{x}) = 0$$

for each $\mathbf{x} \in \mathbf{Z}$. From the first property we see that $\mathbf{x}_0$ is strongly positive. Indeed, $x_0^i = (\mathbf{x}_0|\mathbf{e}_i) \geq |\mathbf{x}_0|^2 > 0$ holds for each $i \in \{1, 2, \ldots n\}$. From the second condition we see that $\mathbf{x}_0$ is orthogonal to $\mathbf{Z}$. Applying Theorem 3.6 we obtain a basis of $\mathbf{Z}^\perp$ consisting of strongly positive vectors. $\square$

As a corollary we obtain a result on the extension of strongly positive operators defined on subspaces of $\mathbb{R}^n$.

Corollary 3.26 *Let $\mathbf{M}$ be a linear subspace of $\mathbb{R}^n$ and $\pi : \mathbf{M} \to \mathbb{R}$ a strongly positive linear functional. Then we can extend π to a strongly positive linear functional on the whole of $\mathbb{R}^n$.*
If $\mathbf{M}$ is a proper subspace, then there exist infinitely many strongly positive extensions. Moreover, the set of strongly positive extensions is convex.

Proof If π is strongly positive we have in particular that $\pi(\mathbf{x}) > 0$ for $\mathbf{x} > 0$. Setting

$$\mathbf{Z} \stackrel{\text{def}}{=} N(\pi)$$

this implies that

$$\mathbf{Z} \cap \mathbb{R}^n_+ = \{0\} \; .$$

Since $\mathbf{Z}$ is $k-1$ dimensional, we find by the above proposition a basis $\mathbf{z}_1, \ldots, \mathbf{z}_{n-k+1}$ of $\mathbf{Z}^\perp$ of strongly positive vectors. Each of these vectors gives rise to a strongly positive extension via Lemma 3.24.
Now if $\mathbf{M} = \mathbb{R}^n$, the dimension of $\mathbf{Z}^\perp$ is 1, and there is only one extension which is obviously π itself. If $\mathbf{M}$ is proper, then $\mathbf{Z}^\perp$ will have dimension $n - k + 1 \geq 2$. So

at least $n - k + 1$ different strongly positive extensions will exist. If $\tilde{\pi}_1$ and $\tilde{\pi}_2$ are two different strongly positive extensions of π and $\lambda \in [0, 1]$, it is easy to see that $\lambda \tilde{\pi}_1 + (1 - \lambda) \tilde{\pi}_2$ is also a strongly positive extension of π. This proves convexity of the set of strongly positive extensions of π and its infinity if it is not a singleton. $\qquad\square$

3.5 Optimal Positive Extensions*

We will now turn to a problem which will arise in a natural way when discussing incomplete markets later on. This material will only be used in that instance and can be skipped in a first reading[1].

3.5.1 The Optimal Extension Problem

Let $\mathbf{M}$ be a proper linear subspace of $\mathbb{R}^n$ of dimension $k < n$. Suppose we are given a strongly positive linear functional $\pi : \mathbf{M} \to \mathbb{R}$. We then know by Corollary 3.26 that there are infinitely many strongly positive extensions of π to the whole space $\mathbb{R}^n$. Hence, both sets

$$\mathcal{E}_\pi^+ \stackrel{\text{def}}{=} \{\psi : \mathbb{R}^n \to \mathbb{R}; \ \psi \text{ is a positive extension of } \pi\}$$

and

$$\mathcal{E}_\pi^{++} \stackrel{\text{def}}{=} \{\psi \in \mathcal{E}_\pi^+; \psi \text{ is a strongly positive extension of } \pi\}$$

are not empty. Moreover,

$$\mathcal{E}_\pi^{++} \subset \mathcal{E}_\pi^+ .$$

Given a fixed $\hat{\mathbf{x}} \in \mathbb{R}^n$ we would like to investigate whether we can choose a positive extension ψ^* of π which maximizes the value of $\hat{\mathbf{x}}$, i.e. such that

$$\psi^*(\hat{\mathbf{x}}) = \sup_{\psi \in \mathcal{E}_\pi^+} \psi(\hat{\mathbf{x}}) \tag{3.5}$$

holds. Of course we first need to know if this supremum is finite.

Lemma 3.27 *Assume there exists a* $\mathbf{y} \in \mathbf{M}$ *with* $\mathbf{y} \geq \hat{\mathbf{x}}$. *Then*

$$\sup_{\psi \in \mathcal{E}_\pi^+} \psi(\hat{\mathbf{x}}) < \infty .$$

Proof For any positive extension ψ of π we have

$$\psi(\hat{\mathbf{x}}) \leq \psi(\mathbf{y}) = \pi(\mathbf{y}) < \infty ,$$

from which the assertion follows. $\qquad\square$

---

[1] All such subsections will be indicated by a star such as in this one

The set

$$\mathcal{D}_{\hat{\mathbf{x}}} \stackrel{\text{def}}{=} \{\mathbf{y} \in \mathbf{M}; \mathbf{y} \geq \hat{\mathbf{x}}\}$$

is the set of vectors in $\mathbf{M}$ which *dominate* $\hat{\mathbf{x}}$. In general this set may be empty or not. However, we have the following result.

Lemma 3.28 $\mathcal{D}_{\hat{\mathbf{x}}} \neq \emptyset$ *for all* $\hat{\mathbf{x}} \in \mathbb{R}^n$ *if and only if* $\mathbf{M}$ *contains a strongly positive vector.*

Proof Assume $\mathbf{M}$ contains a strongly positive vector $\mathbf{y}$. Then $y_i > 0$ for all $i = 1, \ldots, n$. Therefore, we can find $\lambda > 0$ such that $\lambda y_i > \hat{x}_i$ holds for all $i = 1, \ldots, n$. It follows that $\lambda \mathbf{y}$ is a vector in $\mathbf{M}$ such that $\mathbf{y} \geq \hat{\mathbf{x}}$ and $\mathcal{D}_{\hat{\mathbf{x}}} \neq \emptyset$. Assume now that $\mathcal{D}_{\hat{\mathbf{x}}} \neq \emptyset$ for all $\hat{\mathbf{x}} \in \mathbb{R}^n$. In particular we can find a vector $\mathbf{y}$ in $\mathbf{M}$ dominating the strongly positive vector $(1, \ldots, 1) \in \mathbb{R}^n$. It immediately follows that $\mathbf{y}$ is also strongly positive.   □

If $\mathbf{M}$ contains a strongly positive vector, we have that

$$\mathcal{D}_{\hat{\mathbf{x}}} \neq \emptyset$$

for all $\hat{\mathbf{x}} \in \mathbb{R}^n$. The existence of a strongly positive vector together with Theorem 3.6 also imply the following result.

Lemma 3.29 *Assume* $\mathbf{M}$ *contains a strongly positive vector. Then*

$$\mathbf{M} = \mathbf{M}_+ - \mathbf{M}_+ ,$$

i.e. for every $\mathbf{x} \in \mathbf{M}$ *we can find positive vectors* $\mathbf{y}, \mathbf{z} \in \mathbf{M}$ *such that* $\mathbf{x} = \mathbf{y} - \mathbf{z}$.

Proof A basis $\{\mathbf{f}_1, \ldots, \mathbf{f}_k\}$ of $\mathbf{M}$ consisting of strongly positive elements exists by Theorem 3.6. Now, let $\mathbf{x} \in \mathbf{M}$ be arbitrary. Then, we can write it as

$$\mathbf{x} = \sum_{i=1}^{k} \lambda_i \mathbf{f}_i .$$

Define

$$\mathbf{y} \stackrel{\text{def}}{=} \sum_{\substack{1 \leq i \leq k \\ \lambda_i \geq 0}} \lambda_i \mathbf{f}_i \qquad \text{and} \qquad \mathbf{z} \stackrel{\text{def}}{=} - \sum_{\substack{1 \leq i \leq k \\ \lambda_i < 0}} \lambda_i \mathbf{f}_i .$$

Then, obviously, $\mathbf{y}, \mathbf{z}$ are positive elements of $\mathbf{M}$ and $\mathbf{x} = \mathbf{y} - \mathbf{z}$.   □

When investigating the above maximization problem we will also look at a related minimization problem which reads: find $\mathbf{y}^* \in \mathcal{D}_{\hat{\mathbf{x}}}$ such that

$$\pi(\mathbf{y}^*) = \inf_{\mathbf{y} \in \mathcal{D}_{\hat{\mathbf{x}}}} \pi(\mathbf{y}) . \tag{3.6}$$

Lemma 3.30 *Assume* $\mathbf{M}$ *contains a strongly positive vector. Then, for all* $\hat{\mathbf{x}} \in \mathbb{R}^n$,

$$-\infty < \sup_{\psi \in \mathcal{E}_\pi^+} \psi(\hat{\mathbf{x}}) \leq \inf_{\mathbf{y} \in \mathcal{D}_{\hat{\mathbf{x}}}} \pi(\mathbf{y}) < \infty . \tag{3.7}$$

Proof Note that for all $\mathbf{y} \in \mathcal{D}_{\hat{\mathbf{x}}}$ and any positive extension ψ of π we have

$$\psi(\hat{\mathbf{x}}) \leq \psi(\mathbf{y}) = \pi(\mathbf{y}) .$$

Taking the supremum over $\psi \in \mathcal{E}_\pi^+$ we see that

$$-\infty < \sup_{\psi \in \mathcal{E}_\pi^+} \psi(\hat{\mathbf{x}}) \leq \pi(\mathbf{y})$$

for all $\mathbf{y} \in \mathcal{D}_{\hat{\mathbf{x}}}$. Taking now the infimum over $\mathbf{y} \in \mathcal{D}_{\hat{\mathbf{x}}}$ we get (3.7). $\square$

The relationship between (3.5) and (3.6) becomes apparent from the main result of this section. It states that the above supremum and infimum coincide and that both are in fact attained.

Theorem 3.31 *Assume* $\mathbf{M}$ *contains a strongly positive vector. Then, for all* $\hat{\mathbf{x}} \in \mathbb{R}^n$ *there exist* $\psi^* \in \mathcal{E}_\pi^+$ *and* $\mathbf{y}^* \in \mathcal{D}_{\hat{\mathbf{x}}}$ *such that*

$$\psi^*(\hat{\mathbf{x}}) = \sup_{\psi \in \mathcal{E}_\pi^+} \psi(\hat{\mathbf{x}}) , \quad \text{and} \quad \pi(\mathbf{y}^*) = \inf_{\mathbf{y} \in \mathcal{D}_{\hat{\mathbf{x}}}} \pi(\mathbf{y}) .$$

Moreover,

$$\sup_{\psi \in \mathcal{E}_\pi^+} \psi(\hat{\mathbf{x}}) = \inf_{\mathbf{y} \in \mathcal{D}_{\hat{\mathbf{x}}}} \pi(\mathbf{y}) .$$

For the proof of this theorem we need a simple result from linear programming theory which we develop in the next section. But before proceeding let us make a few remarks.

The next example shows that while it may be possible to find a maximizing positive extension ψ^*, a maximizing strongly positive extension may fail to exist.

Example 3.32 *In* $\mathbb{R}^2$ *consider the one-dimensional subspace*

$$\mathbf{M} \overset{def}{=} \{\lambda \mathbf{f}; \lambda \in \mathbb{R}\} ,$$

where $\mathbf{f} \overset{def}{=} (1,1)$, *which is obviously strongly positive. Define now the strongly positive linear functional* $\pi : \mathbf{M} \to \mathbb{R}$ *by*

$$\pi(\lambda \mathbf{f}) \overset{def}{=} a\lambda ,$$

for some $a > 0$. *Denote the standard basis vectors by* $\mathbf{e}_1 = (0,1)$ *and* $\mathbf{e}_2 = (1,0)$ *and note that*

$$\mathbf{f} = \mathbf{e}_1 + \mathbf{e}_2 .$$

*It is easy to see that if ψ is an extension of π to the whole of $\mathbb{R}^2$, then there exists
a real number b such that*

$$\psi = \psi_b(x_1, x_2) \overset{def}{=} bx_1 + (a-b)x_2 ,$$

for all $\mathbf{x} = (x_1, x_2)$. In particular the set of all positive extensions is given by

$$\mathcal{E}_\pi^+ = \{\psi_b; a \ge b \ge 0\} .$$

Take now $\hat{\mathbf{x}} \overset{def}{=} \mathbf{e}_2$. For $\psi_b \in \mathcal{E}_\pi^+$ we have

$$\psi_b(\hat{\mathbf{x}}) = a - b \ge 0 .$$

It follows that

$$\sup_{\psi \in \mathcal{E}_\pi^+} \psi(\hat{\mathbf{x}}) = \sup_{0 \le b \le a} \psi_b(\hat{\mathbf{x}}) = \psi_0(\hat{\mathbf{x}}) = a .$$

Since $\psi_0(\mathbf{e}_1) = 0$ holds, we see that ψ_0 is positive but not strongly positive.

However, as the following remark shows, we can always approximate an optimal
positive extension by a sequence of strongly positive extensions.

Remark 3.33 *Let ψ_k be a minimizing sequence in $\mathcal{E}_\pi^+$, i.e.*

$$\lim_{k \to \infty} \psi_k(\hat{x}) = \sup_{\psi \in \mathcal{E}_\pi^+} \psi(\hat{x}) .$$

*Take any strongly positive extension $\tilde{\psi}$ of π and a strictly positive sequence $\lambda_k \nearrow 1$.
Then it is easy to verify that*

$$\tilde{\psi}_k \overset{def}{=} \lambda_k \psi_k + (1 - \lambda_k)\tilde{\psi}$$

is a strongly positive extension of π. It follows that

$$\lim_{k \to \infty} \tilde{\psi}_k(\hat{x}) = \lim_{k \to \infty} \psi_k(\hat{x}) = \sup_{\psi \in \mathcal{E}_\pi^+} \psi(\hat{x}) .$$

Hence, we always have

$$\sup_{\psi \in \mathcal{E}_\pi^{++}} \psi(\hat{\mathbf{x}}) = \sup_{\psi \in \mathcal{E}_\pi^+} \psi(\hat{\mathbf{x}}) .$$

3.5.2   A Linear Programming Problem

Let $\mathbf{M}$ and $\mathbf{Y}$ be linear subspaces of $\mathbb{R}^n$ and $\mathbb{R}^m$, respectively. Consider the following minimization problem.

- Minimize $\nu(\mathbf{x})$,

- subject to the constraint $\mathbf{x} \in \mathbf{K}_{A,\mathbf{b}} = \{\mathbf{x} \in \mathbf{M}; \mathbf{x} \ge 0 \text{ and } A\mathbf{x} = \mathbf{b}\}$,

where $\nu : \mathbf{M} \to \mathbb{R}$ is a linear functional, $A : \mathbf{M} \to \mathbf{Y}$ is a linear mapping and
$\mathbf{b} \in \mathbf{Y}$ is a vector.

This is an example of a so-called *linear programming* problem. We shall denote it
by $LP(\nu, A, \mathbf{b})$. The set $\mathbf{K}_{A,\mathbf{b}}$ is called the *feasible* region and the functional ν the
objective function.

Existence of solutions

The next result states that if the feasible region is not empty and the objective function is bounded from below on the feasible region, there always exists a solution to $LP(\nu, A, \mathbf{b})$.

Theorem 3.34 *Assume that* $\mathbf{K}_{A,\mathbf{b}} \neq \emptyset$ *and that*

$$\lambda_0 \stackrel{def}{=} \inf_{\mathbf{x} \in \mathbf{K}_{A,\mathbf{b}}} \nu(\mathbf{x}) > -\infty \,,$$

i.e. ν *is bounded from below on* $\mathbf{K}_{A,\mathbf{b}}$. *Then there exists an* $\mathbf{x}^* \in \mathbf{K}_{A,\mathbf{b}}$ *such that*

$$\nu(\mathbf{x}^*) = \inf_{\mathbf{x} \in \mathbf{K}_{A,\mathbf{b}}} \nu(\mathbf{x}) \,.$$

Proof Note that by definition of λ_0

$$\lambda_0 \leq \nu(\mathbf{x}) \tag{3.8}$$

holds for all $\mathbf{x} \in \mathbf{K}_{A,\mathbf{b}}$. Define a linear operator $\tilde{A} : \mathbf{M} \to \mathbf{Y} \times \mathbb{R}$ by

$$\tilde{A}\mathbf{x} \stackrel{def}{=} (A\mathbf{x}, \nu(\mathbf{x})) \,.$$

If we can show that the system

$$\tilde{A}\mathbf{x} = \tilde{\mathbf{b}} \stackrel{def}{=} (\mathbf{b}, \lambda_0) \tag{3.9}$$

admits a positive solution $\mathbf{x}^*$ we will have proved the theorem since then $A\mathbf{x}^* = \mathbf{b}$ and $\nu(\mathbf{x}^*) = \lambda_0$. We proceed to show the existence of such an $\mathbf{x}^*$ by applying Farkas' Lemma.

Let thus $\tilde{\mathbf{y}} = (\mathbf{y}, \lambda) \in \mathbf{Y} \times \mathbb{R}$ be such that

$$0 \leq (\tilde{A}\mathbf{x}|\tilde{\mathbf{y}}) = (A\mathbf{x}|\mathbf{y}) + \lambda \cdot \nu(\mathbf{x}) \tag{3.10}$$

holds for all positive $\mathbf{x}$ in $\mathbb{R}^n$. If we can show that

$$0 \leq (\tilde{\mathbf{b}}|\tilde{\mathbf{y}}) = (\mathbf{b}|\mathbf{y}) + \lambda \cdot \lambda_0 \,, \tag{3.11}$$

holds, then by Farkas' Lemma, the system (3.9) will admit a positive solution $\mathbf{x}^*$. Inequality (3.10) holds in particular for all $\hat{\mathbf{x}} \in K_{A,\mathbf{b}} \neq \emptyset$, i.e. for all $\hat{\mathbf{x}} \geq 0$ with $\mathbf{b} = A\hat{\mathbf{x}}$. Hence,

$$0 \leq (\mathbf{b}|\mathbf{y}) + \lambda \cdot \nu(\hat{\mathbf{x}}) \,. \tag{3.12}$$

In order to show that (3.11) holds, we distinguish two cases.

Case I: $\lambda \geq 0$. From (3.12) we infer that

$$0 \leq (\mathbf{b}|\mathbf{y}) + \lambda \cdot \inf_{\hat{\mathbf{x}} \in \mathbf{K}_{A,\mathbf{b}}} \nu(\hat{\mathbf{x}}) = (\mathbf{b}|\mathbf{y}) + \lambda \cdot \lambda_0 \,.$$

In order to prove the above proposition we will apply our linear programming existence result to a slightly modified problem and invoke the simple principle of Lemma 3.36. The space

$$\mathcal{M} \overset{\text{def}}{=} \mathbf{M} \times \mathbf{M} \times \mathbb{R}^n$$

is a linear subspace of $\mathbb{R}^n \times \mathbb{R}^n \times \mathbb{R}^n$. Define a linear functional $\nu : \mathcal{M} \to \mathbb{R}$, a linear operator $A : \mathcal{M} \to \mathbb{R}^n$, and a vector $\mathbf{b} \in \mathbb{R}^n$ by setting for any $(\mathbf{u}, \mathbf{v}, \mathbf{z}) \in \mathcal{M}$

$$\nu(\mathbf{u}, \mathbf{v}, \mathbf{z}) \overset{\text{def}}{=} \pi(\mathbf{u}) - \pi(\mathbf{v}) = \pi(\mathbf{u} - \mathbf{v}) ,$$

$$A(\mathbf{u}, \mathbf{v}, \mathbf{z}) \overset{\text{def}}{=} \mathbf{u} - \mathbf{v} - \mathbf{z} .$$

Define now the set

$$\mathbf{K}_{A,\hat{\mathbf{x}}} \overset{\text{def}}{=} \{(\mathbf{u}, \mathbf{v}, \mathbf{z}) \in \mathcal{M}; A(\mathbf{u}, \mathbf{v}, \mathbf{z}) = \hat{\mathbf{x}} \text{ and } (\mathbf{u}, \mathbf{v}, \mathbf{z}) \geq 0\} .$$

Below we will show that there exists a $(\mathbf{u}^*, \mathbf{v}^*, \mathbf{z}^*) \in \mathbf{K}_{A,\hat{\mathbf{x}}}$ such that

$$\nu(\mathbf{u}^*, \mathbf{v}^*, \mathbf{z}^*) = \min_{(\mathbf{u},\mathbf{v},\mathbf{z}) \in \mathbf{K}_{A,\hat{\mathbf{x}}}} \nu(\mathbf{u}, \mathbf{v}, \mathbf{z}) . \tag{3.14}$$

The following lemma in conjunction with Lemma 3.36 then implies the statement of Proposition 3.37.

Lemma 3.38 *Define a linear map $F : \mathcal{M} \to \mathbb{R}^n$ by setting*

$$F(\mathbf{u}, \mathbf{v}, \mathbf{z}) \overset{\text{def}}{=} \mathbf{u} - \mathbf{v} .$$

The following statements hold:

   a) $\nu(\mathbf{u}, \mathbf{v}, \mathbf{z}) = \pi(F(\mathbf{u}, \mathbf{v}, \mathbf{z}))$; and

   b) $F : \mathbf{K}_{A,\hat{\mathbf{x}}} \to \mathcal{D}_{\hat{\mathbf{x}}}$ is surjective.

In particular, since $\mathcal{D}_{\hat{\mathbf{x}}} \neq \emptyset$, we have that $\mathbf{K}_{A,\hat{\mathbf{x}}} \neq \emptyset$.

Proof a) This follows immediately from

$$\nu(\mathbf{u}, \mathbf{v}, \mathbf{z}) = \pi(\mathbf{u} - \mathbf{v}) = \pi(F(\mathbf{u}, \mathbf{v}, \mathbf{z})) .$$

b) We start by showing that F maps $\mathbf{K}_{A,\hat{\mathbf{x}}}$ into $\mathcal{D}_{\hat{\mathbf{x}}}$. To do this let $(\mathbf{u}, \mathbf{v}, \mathbf{z}) \in \mathbf{K}_{A,\hat{\mathbf{x}}}$ be given. Thus,

$$A(\mathbf{u}, \mathbf{v}, \mathbf{z}) = \mathbf{u} - \mathbf{v} - \mathbf{z} = \hat{\mathbf{x}} ,$$

so that

$$F(\mathbf{u}, \mathbf{v}, \mathbf{z}) = \mathbf{u} - \mathbf{v} = \hat{\mathbf{x}} + \mathbf{z} .$$

Since $\mathbf{z} \geq 0$ we get $\hat{\mathbf{x}} \leq F(\mathbf{u}, \mathbf{v}, \mathbf{z})$. Moreover,

$$F(\mathbf{u}, \mathbf{v}, \mathbf{z}) = \mathbf{u} - \mathbf{v} \in \mathbf{M} \ ,$$

since both $\mathbf{u}$ and $\mathbf{v}$ belong to $\mathbf{M}$. It follows that $F(\mathbf{u}, \mathbf{v}, \mathbf{z}) \in \mathcal{D}_{\hat{\mathbf{x}}}$. We next show that F is surjective. Indeed, by Lemma 3.29, given $\mathbf{y} \in \mathcal{D}_{\hat{\mathbf{x}}}$ we can find positive vectors $\mathbf{u}, \mathbf{v} \in \mathbf{M}$ such that

$$\mathbf{y} = \mathbf{u} - \mathbf{v} \ ,$$

in particular

$$F(\mathbf{u}, \mathbf{v}, \mathbf{z}) = \mathbf{y}$$

holds for all $\mathbf{z} \in \mathbb{R}^n$. Set furthermore,

$$\mathbf{z} \stackrel{\text{def}}{=} \mathbf{y} - \hat{\mathbf{x}} \ .$$

Since $\mathbf{y} \geq \hat{\mathbf{x}}$ we see that $\mathbf{z} \geq 0$. Since

$$A(\mathbf{u}, \mathbf{v}, \mathbf{z}) = \mathbf{u} - \mathbf{v} - \mathbf{z} = \mathbf{y} - (\mathbf{y} - \hat{\mathbf{x}}) = \hat{\mathbf{x}} \ ,$$

it immediately follows that $(\mathbf{u}, \mathbf{v}, \mathbf{z}) \in \mathbf{K}_{A,\hat{\mathbf{x}}}$.   □

From this result and Lemma 3.36 it follows that

$$-\infty < \mu_{\hat{\mathbf{x}}} = \inf_{\mathbf{y} \in \mathcal{D}_{\hat{\mathbf{x}}}} \pi(\mathbf{y}) = \inf_{(\mathbf{u}, \mathbf{v}, \mathbf{z}) \in \mathbf{K}_{A,\hat{\mathbf{x}}}} \nu(\mathbf{u}, \mathbf{v}, \mathbf{z}) \ , \tag{3.15}$$

so that ν is bounded from below. Since we have also shown that $\mathbf{K}_{A,\hat{\mathbf{x}}} \neq \emptyset$, applying Theorem 3.34 we obtain the existence of $(\mathbf{u}^*, \mathbf{v}^*, \mathbf{z}^*) \in \mathbf{K}_{A,\hat{\mathbf{x}}}$ satisfying (3.14). It now follows from Lemma 3.36 that $\mathbf{y}^* \stackrel{\text{def}}{=} F(\mathbf{u}^*, \mathbf{v}^*, \mathbf{z}^*)$ satisfies (3.13) concluding the proof of the proposition.

Proof of Step II: $\sigma_{\hat{\mathbf{x}}} = \mu_{\hat{\mathbf{x}}}$

We shall now prove the equality of the infimum of (3.6) and the supremum of (3.5).

Proposition 3.39 *We have*

$$\sigma_{\hat{\mathbf{x}}} = \mu_{\hat{\mathbf{x}}} \ . \tag{3.16}$$

We continue to use the notation of Step I. We shall prove the existence of an element $(\mathbf{u}^*, \mathbf{v}^*, \mathbf{z}^*) \in \mathbf{K}_{A,\hat{\mathbf{x}}}$ such that

$$\nu(\mathbf{u}^*, \mathbf{v}^*, \mathbf{z}^*) = \sigma_{\hat{\mathbf{x}}} \ . \tag{3.17}$$

Then, using this together with Lemma 3.30 and equation 3.15 we get

$$\sigma_{\hat{\mathbf{x}}} \leq \mu_{\hat{\mathbf{x}}} = \inf_{(\mathbf{u},\mathbf{v},\mathbf{z}) \in \mathbf{K}_{A,\hat{\mathbf{x}}}} \nu(\mathbf{u},\mathbf{v},\mathbf{z}) \leq \nu(\mathbf{u}^*,\mathbf{v}^*,\mathbf{z}^*) = \sigma_{\hat{\mathbf{x}}} \ .$$

This obviously implies $\sigma_{\hat{\mathbf{x}}} = \mu_{\hat{\mathbf{x}}}$, thus establishing the proposition.

We will prove (3.17) by applying Farkas' Lemma in a manner very similar to the proof of Theorem 3.34. To that effect we introduce a linear operator $\tilde{A} : \mathcal{M} \rightarrow \mathbb{R}^n \times \mathbb{R}$ and a vector $\tilde{\mathbf{b}} \in \mathbb{R}^n \times \mathbb{R}$ by setting

$$\tilde{A}(\mathbf{u},\mathbf{v},\mathbf{z}) \stackrel{\text{def}}{=} (A(\mathbf{u},\mathbf{v},\mathbf{z}), \nu(\mathbf{u},\mathbf{v},\mathbf{z})) = (\mathbf{u} - \mathbf{v} - \mathbf{z}, \pi(\mathbf{u} - \mathbf{v})),$$
$$\tilde{\mathbf{b}} \stackrel{\text{def}}{=} (\hat{\mathbf{x}}, \sigma_{\hat{\mathbf{x}}}) \ .$$

For $(\mathbf{u}^*, \mathbf{v}^*, \mathbf{z}^*) \in \mathcal{M}$, belonging to $\mathbf{K}_{A,\hat{\mathbf{x}}}$ and satisfying (3.17) is equivalent to being a positive solution of

$$\tilde{A}(\mathbf{u}^*, \mathbf{v}^*, \mathbf{z}^*) = \tilde{\mathbf{b}} \ . \tag{3.18}$$

We proceed to prove the existence of a positive solution of (3.18). Assume $\tilde{\mathbf{y}} = (\mathbf{c}, \lambda) \in \mathbb{R}^n \times \mathbb{R}$ is given such that

$$0 \leq (\tilde{A}(\mathbf{u},\mathbf{v},\mathbf{z})|\tilde{\mathbf{y}}) = (\mathbf{u} - \mathbf{v} - \mathbf{z}|\mathbf{c}) + \lambda \cdot \pi(\mathbf{u} - \mathbf{v}) \tag{3.19}$$

holds for all positive $(\mathbf{u},\mathbf{v},\mathbf{z}) \in \mathcal{M}$. If we can show that

$$0 \leq (\tilde{\mathbf{b}}|\tilde{\mathbf{y}}) = (\hat{\mathbf{x}}|\mathbf{c}) + \lambda \cdot \sigma_{\hat{\mathbf{x}}} \tag{3.20}$$

also holds, we can invoke Farkas' Lemma to obtain a positive solution of (3.18). We will need the following simple implication of (3.19).

Lemma 3.40 *If (3.19) is satisfied, then* $\phi : \mathbb{R}^n \rightarrow \mathbb{R}$ *given by*

$$\phi(\mathbf{x}) \stackrel{\text{def}}{=} -\frac{1}{\lambda}(\mathbf{x}|\mathbf{c})$$

for all $\mathbf{x} \in \mathbb{R}^n$, *defines a positive extension of* π, *i.e.* $\phi \in \mathcal{E}_\pi^+$.

Proof Choosing $\mathbf{z} = 0$ in (3.19) and using, from Lemma 3.29, that every $\mathbf{y} \in \mathbf{M}$ can be written as $\mathbf{y} = \mathbf{u} - \mathbf{v}$ with positive $\mathbf{u}, \mathbf{v} \in \mathbf{M}$, we immediately get the inequality

$$0 \leq (\mathbf{y}|\mathbf{c}) + \lambda \cdot \pi(\mathbf{y})$$

for all $\mathbf{y} \in \mathbf{M}$. Since $\mathbf{M}$ is a linear subspace, with $\mathbf{y}$ it also contains $-\mathbf{y}$. Therefore, we get that

$$0 = (\mathbf{y}|\mathbf{c}) + \lambda \cdot \pi(\mathbf{y})$$

holds for all $\mathbf{y} \in \mathbf{M}$. From this we infer that

$$\pi(\mathbf{y}) = -\frac{1}{\lambda}(\mathbf{y}|\mathbf{c}) = \phi(\mathbf{y})$$

for all $\mathbf{y} \in \mathbf{M}$, so that ϕ is an extension of π. To see that it is a positive extension note that choosing $\mathbf{u} = \mathbf{v} = 0$ and $\mathbf{z} \geq 0$ we get from (3.19)

$$0 \leq -\frac{1}{\lambda}(\mathbf{z}|\mathbf{c}) = \phi(\mathbf{z}) \ .$$

$\square$

In order to prove (3.20) we distinguish two cases.

Case I: $\lambda \leq 0$. Taking $(\mathbf{u}, \mathbf{v}, \mathbf{z}) \in \mathbf{K}_{A,\hat{\mathbf{x}}}$ and noting that in this case $\hat{\mathbf{x}} \leq \mathbf{u} - \mathbf{v}$, i.e. $\mathbf{u} - \mathbf{v} \in \mathcal{D}_{\hat{\mathbf{x}}}$, we have by Lemma 3.30,

$$\sigma_{\hat{\mathbf{x}}} = \sup_{\psi \in \mathcal{E}_\pi^+} \psi(\hat{\mathbf{x}}) \leq \mu_{\hat{\mathbf{x}}} = \inf_{\mathbf{y} \in \mathcal{D}_{\hat{\mathbf{x}}}} \pi(\mathbf{y}) \leq \pi(\mathbf{u} - \mathbf{v}) \ .$$

It follows since $\lambda \leq 0$ that

$$\lambda \pi(\mathbf{u} - \mathbf{v}) \leq \lambda \sigma_{\hat{\mathbf{x}}} \ .$$

This, together with $\mathbf{u} - \mathbf{v} - \mathbf{z} = \hat{\mathbf{x}}$, and (3.19) yields

$$0 \leq (\hat{\mathbf{x}}|\mathbf{c}) + \lambda \cdot \pi(\mathbf{u} - \mathbf{v}) \leq (\hat{\mathbf{x}}|\mathbf{c}) + \lambda \sigma_{\hat{\mathbf{x}}} \ ,$$

which is (3.20) in case $\lambda \leq 0$.

Case II: $\lambda > 0$. From the above lemma we get

$$-\frac{1}{\lambda}(\hat{\mathbf{x}}|\mathbf{c}) = \phi(\hat{\mathbf{x}}) \leq \sup_{\psi \in \mathcal{E}_\pi^+} \psi(\hat{\mathbf{x}}) = \sigma_{\hat{\mathbf{x}}} \ .$$

Since $\lambda > 0$ this immediately implies

$$0 \leq (\hat{\mathbf{x}}|\mathbf{c}) + \lambda \sigma_{\hat{\mathbf{x}}} \ ,$$

proving (3.20) in case $\lambda > 0$. This completes the proof of Step II.

Proof of Step III: Existence of ψ^*

We proceed to prove that the supremum $\sigma_{\hat{\mathbf{x}}}$ is achieved.

Proposition 3.41 *There exists a $\psi^* \in \mathcal{E}_\pi^+$ with*

$$\psi^*(\hat{\mathbf{x}}) = \sup_{\psi \in \mathcal{E}_\pi^+} \psi(\hat{\mathbf{x}}) = \mu_{\hat{\mathbf{x}}} \ . \tag{3.21}$$

Here we will again apply the linear programming existence result to a suitable problem which will give us via Lemma 3.36 the desired existence of ψ^*. For greater clarity define $f : \mathcal{E}_\pi^+ \to \mathbb{R}$ by

$$f(\psi) \stackrel{\text{def}}{=} \psi(\hat{\mathbf{x}}) \ .$$

Then, the proposition can be equivalently formulated as asserting the existence of $\psi^* \in \mathcal{E}_\pi^+$ such that

$$\psi^*(\hat{\mathbf{x}}) = f(\psi^*) = \sup_{\psi \in \mathcal{E}_\pi^+} f(\psi) \ . \tag{3.22}$$

Recall that $dim(\mathbf{M}) = k$. Let $\mathbf{f}_1, \mathbf{f}_2, \ldots, \mathbf{f}_k$ be a basis of $\mathbf{M}$ and define $A : \mathbb{R}^n \to \mathbb{R}^k$ and $\nu : \mathbb{R}^n \to \mathbb{R}$ by setting

$$A\mathbf{z} \stackrel{\text{def}}{=} ((\mathbf{f}_1|\mathbf{z}), \ldots (\mathbf{f}_k|\mathbf{z})) \ ,$$

$$\nu(\mathbf{z}) \stackrel{\text{def}}{=} (\hat{\mathbf{x}}|\mathbf{z}) \ .$$

Finally, define $\mathbf{b} \in \mathbb{R}^k$ by

$$\mathbf{b} \stackrel{\text{def}}{=} (\pi(\mathbf{f}_1), \ldots, \pi(\mathbf{f}_k)) \ .$$

Consider the set

$$\mathbf{K}_{A,\mathbf{b}} \stackrel{\text{def}}{=} \{\mathbf{z} \in \mathbb{R}^n; A\mathbf{z} = \mathbf{b} \text{ and } \mathbf{z} \geq 0\} \ .$$

Below we will show that there exists a $\mathbf{z}^* \in \mathbf{K}_{A,\mathbf{b}}$ such that

$$\nu(\mathbf{z}^*) = \sup_{\mathbf{z} \in \mathbf{K}_{A,\mathbf{b}}} \nu(\mathbf{z}) \ . \tag{3.23}$$

As in Step I, we now prove a lemma which together with Lemma 3.36 will establish the existence of a solution of (3.22), thus proving Proposition 3.41.

Recall that by the results in Section 3.1.1 $\pi_{\mathbf{z}}(\mathbf{x}) \stackrel{\text{def}}{=} (\mathbf{x}|\mathbf{z})$ defines a linear functional $\mathbb{R}^n \to \mathbb{R}$ for every $\mathbf{z} \in \mathbb{R}^n$. Moreover, for each linear functional $\psi : \mathbb{R}^n \to \mathbb{R}$ there exists a unique $\mathbf{z} \in \mathbb{R}^n$ such that $\psi = \pi_{\mathbf{z}}$. By Corollary 3.10 we know that $\pi_{\mathbf{z}}$ is (strongly) positive if and only if $\mathbf{z}$ is (strongly) positive.

Lemma 3.42 *For each $\mathbf{z} \in \mathbf{K}_{A,\mathbf{b}}$ we have $\pi_{\mathbf{z}} \in \mathcal{E}_\pi^+$. Moreover, setting*

$$F(\mathbf{z}) \stackrel{\text{def}}{=} \pi_{\mathbf{z}} \ ,$$

we have

  a) $\nu(\mathbf{z}) = f(F(\mathbf{z}))$; and

  b) $F : \mathbf{K}_{A,\mathbf{b}} \to \mathcal{E}_\pi^+$ is a surjection. In particular, since $\mathcal{E}_\pi^+ \neq \emptyset$, we have that $\mathbf{K}_{A,\mathbf{b}} \neq \emptyset$.

Proof

a) This follows since by definition of f

$$\nu(\mathbf{z}) = (\hat{\mathbf{x}}|\mathbf{z}) = \pi_{\mathbf{z}}(\hat{\mathbf{x}}) = f(\pi_{\mathbf{z}}) = f(F(\mathbf{z})) \ .$$

b)We first prove that F maps $\mathbf{K}_{A,\mathbf{b}}$ into $\mathcal{E}_{\pi}^{+}$. Take any $\mathbf{z} \in \mathbf{K}_{A,\mathbf{b}}$. Then, since F preserves positivity we have $F(\mathbf{z}) \geq 0$. Moreover, we have

$$((f_1|\mathbf{z}), \ldots, (f_k|\mathbf{z})) = A\mathbf{z} = \mathbf{b} = (\pi(\mathbf{f}_1), \ldots, \pi(\mathbf{f}_k)) \ .$$

Any $\mathbf{y} \in M$ can be written as $\mathbf{y} = c_1 \mathbf{f}_1 + \ldots + c_k \mathbf{f}_k$. Thus,

$$\pi_{\mathbf{z}}(\mathbf{y}) = \sum_{i=1}^{k} c_i F_{\mathbf{z}}(\mathbf{f}_i) = \sum_{i=1}^{k} c_i (\mathbf{f}_i|\mathbf{z}) = \sum_{i=1}^{k} c_i \pi(\mathbf{f}_i) = \pi(\mathbf{y}) \ .$$

It follows that $\pi_{\mathbf{z}}$ is a positive extension of π and thus belongs to $\mathcal{E}_{\hat{\pi}}^{+}$.
To prove that $\mathcal{E}_{\hat{\pi}}^{+} = F(\mathbf{K}_{A,\mathbf{b}})$, take any $\psi \in \mathcal{E}_{\hat{\pi}}^{+}$. By the remarks preceding the Lemma we find a $\mathbf{z} \in \mathbb{R}^n$ such that $\pi_{\mathbf{z}} = \psi$. We show that $\mathbf{z} \in \mathbf{K}_{A,\mathbf{b}}$. Now, since ψ is an extension of π, we find that

$$(\mathbf{f}_j|z) = \pi_{\mathbf{z}}(\mathbf{f}_j) = \psi(\mathbf{f}_j) = \pi(\mathbf{f}_j)$$

for all $j = 1, 2, \ldots, k$. It follows that

$$A\mathbf{z} = ((f_1|\mathbf{z}), \ldots, (f_k|\mathbf{z})) = (\pi(\mathbf{f}_1), \ldots, \pi(\mathbf{f}_k)) = \mathbf{b} \ .$$

Moreover, since ψ is a positive extension we conclude that $\mathbf{z} \geq 0$. It follows that $\mathbf{z} \in \mathbf{K}_{A,\mathbf{b}}$. This proves that F is surjective.   □

From the above result and Lemma 3.36 we get that

$$\sup_{\mathbf{z} \in \mathbf{K}_{A,\mathbf{b}}} \nu(\mathbf{z}) = \sup_{\psi \in \mathcal{E}_{\hat{\pi}}^{+}} f(\psi) = \sigma_{\hat{\mathbf{x}}} < \infty \ . \tag{3.24}$$

This implies that ν is bounded from above on $\mathbf{K}_{A,\hat{\mathbf{b}}}$. Moreover, since we also showed that $\mathbf{K}_{A,\hat{\mathbf{b}}} \neq \emptyset$, applying Corollary 3.35 we obtain the existence of $\mathbf{z}^*$ satisfying (3.23). From Lemma 3.36 it now follows that $F_{\mathbf{z}^*} \stackrel{\text{def}}{=} \psi^*$ satisfies (3.21).

Concluding Remarks and Suggestions for Further Reading

The material of this chapter is important in many branches of mathematical finance and economics. It also provides the basis for translating many of the finance problems treated in this book into the geometric language of linear algebra.

There is a well-developed theory of positive linear operators also for infinite dimensional ordered spaces. A classic reference is [51]. Economic applications can be found in [1] and [2]. References on linear programming, a topic we have barely touched upon here, are [14] or [15].

The next two chapters start developing the probability theory necessary for understanding the models in this book. Readers who are already acquainted with the elements of probability theory may want to skip them or just browse through them to get a feel for our notation.

Chapter 4

Finite Probability Spaces

At any rate, according to the statistical view, the mathematical laws of nature describe at best how nature will probably behave, but they do not preclude that the earth may suddenly wander off into space. Nature can make up her own mind and decide not to do what is most probable.

M. Kline

Before taking up the study of general one-period models involving an arbitrary but finite number of "states" and "securities" we will introduce in this and the next chapter some concepts from probability theory. We shall be concerned only with finite probability spaces and, at this stage, we will only describe the elements of the theory. More advanced aspects will be developed in later chapters.

4.1 Finite Probability Spaces

At the start of probability theory we find the concepts of a sample space and of a probability measure thereon.

4.1.1 Sample Spaces and Events

When modelling a random experiment admitting only a finite number of outcomes we start with the specification of a finite set

$$\Omega = \{\omega_1, \ldots \omega_n\} \, ,$$

called a *sample space*, whose elements correspond to the *possible outcomes* of the random experiment and are usually called the *elementary events*.

Any subset of Ω will be called an *event*. We shall say an event $A \subset \Omega$ has *occurred* if, once the experiment has been performed, its outcome ω belongs to A. Similarly, if the outcome of the experiment is not an element of A, then we say that the event has *not occurred*. The events $\{\omega_i\}$, $i \in \{1, 2, \ldots, n\}$ are usually called the *elementary events*.

Events A and B are said to be *mutually exclusive*, or *incompatible*, if $A \cap B = \emptyset$.

4.1.2   Probability Measures

The next step consists in specifying the probability of occurrence of each event. In other words we need to specify a function which assigns to each event a number which we interpret as its probability of occurrence. Formally speaking, a function[1]

$$P : 2^{\Omega} \to [0, 1]$$

is a *probability measure* or *probability distribution* on Ω if the following two properties are satisfied:

- $P(\Omega) = 1$; and

- If $A_1, \cdots, A_r$ are pairwise disjoint sets, i.e. $A_i \cap A_j = \emptyset$ when $i \neq j$, then

$$P(A_1 \cup \cdots \cup A_r) = \sum_{i=1}^{r} P(A_i) \, .$$

The pair (Ω, P) is called a *(finite) probability space*.

Intuitively, we think of the number $P(A)$ as representing the relative frequency with which A will occur in a large number of repetitions of the random experiment. So if $P(A) = 0.7$ and we perform our experiment 100 times, we will expect to observe A roughly 70 out of the 100 times. The first property is another way of saying that the sample space Ω contains every possible outcome, so that when the random experiment is performed we can be certain to observe Ω. The second property, called *additivity*, just says that if we have a collection of mutually exclusive events, then the odds that at least one of them occurs is just the sum of the odds that each of them occurs.

Probabilities on finite spaces can be easily described . For each outcome ω_i we can set

$$p_i \overset{\text{def}}{=} P(\omega_i) \overset{\text{def}}{=} P(\{\omega_i\}) \, .$$

Any event A can be described by listing its elements[2]

$$A = \{\omega_{i_1}, \omega_{i_2}, \ldots, \omega_{i_l}\} \, ,$$

---

[1] If A is a set we call the collection of all subsets of A the *power set* of A and denote it by 2^A.
[2] The notation $i_1, \ldots, i_r$ is just standard notation used when selecting a subsequence with r elements from a sequence $(i)_{1 \leq i \leq n}$.

and A can be viewed as the disjoint union of the elementary events belonging to it. Therefore, by the additivity property, we have:

$$P(A) \stackrel{\text{def}}{=} p_{i_1} + p_{i_2} + \ldots + p_{i_l}.$$

Because we have required that $P(\Omega) = 1$, we also must have that

$$P(\Omega) = p_1 + p_2 + \ldots + p_n = 1 .$$

In the manner just described any n-tuple $(p_1, p_2, \ldots, p_n)$ satisfying

$$p_1 + p_2 + \ldots + p_n = 1 .$$

uniquely determines a probability measure on Ω. We will sometimes by abuse of notation write $P = (p_1, p_2, \ldots, p_n)$.
We next list some simple but fundamental properties of probability measures.

Proposition 4.1 *For any probability space we have the following properties:*

a) *For events A and B we have[3] $P(A \setminus B) = P(A) - P(A \cap B)$.*

b) *If A and B are events, then $P(A \cup B) = P(A) + P(B) - P(A \cap B)$. In particular, whenever two events A and B are disjoint, i.e. $A \cap B = \emptyset$, we have $P(A \cup B) = P(A) + P(B)$.*

c) *For an event A and its complement A^c we have $P(A^c) = 1 - P(A)$.*

d) *If $A \subset B$ holds, then $P(A) \leq P(B)$.*

Proof To prove the first assertion just note that A is the disjoint union of $A \cap B$ and $A \setminus B = A \cap B^c$. Therefore, by the additivity property of P, we get

$$P(A) = P(A \cap B) + P(A \setminus B) ,$$

proving the assertion.
The second assertion follows from the fact that we can write $A \cup B$ as the disjoint sum of A and $B \setminus A$. Hence, using our first result,

$$P(A \cup B) = P(A) + P(B \setminus A) = P(A) + P(B) - P(A \cap B) .$$

In the same vein we can write Ω as the disjoint union of A and A^c, so that

$$1 = P(\omega) = P(A) + P(A^c) ,$$

proving the third assertion.
Finally, if A is a subset of B we can write B as the disjoint union of A and $B \setminus B$, so that

$$P(A) \leq P(A) + P(B \setminus B) = P(B) .$$

$\square$

---

[3]Recall that $A \setminus B$ is the subset of A containing all elements in A which are not contained in B, i.e. $A \setminus B \stackrel{\text{def}}{=} A \cap B^c$.

4.2   Laplace Experiments

Classical probability theory was concerned with the simplest possible probability spaces. These correspond to the situation where a sample space $\Omega = \{\omega_1, \ldots \omega_N\}$ and a probability measure P are given such that the probability $P(\omega_j)$ of an elementary event $\{\omega_j\}$ occurring is the same for all $j = 1, \ldots, N$. Therefore, there exists a $0 < p < 1$ such that $P(\omega_j) = p$ for all $j = 1, \ldots, N$. From the properties of a probability measure we then get

$$1 = P(\Omega) = P(\omega_1) + \cdots + P(\omega_N) = N \cdot p .$$

It follows that

$$P(\omega_j) = p = \frac{1}{N}$$

holds for all $j = 1, \ldots, N$. If $A \subset \Omega$ is any event then

$$P(A) = \sum_{\omega \in A} P(\{\omega\}) = \sum_{\omega \in A} \frac{1}{N} = \frac{\#A}{N} ,$$

where $\#A$ denotes the *cardinality* (the number of elements) of A.
It is not always easy to determine the cardinality of a given event A. We shall mention some standard combinatorial techniques further below.

Tossing a coin

A classical example consists in the tossing of a fair coin. Here the sample space is naturally given by

$$\Omega \stackrel{\text{def}}{=} \{H, T\}$$

where H stands for "heads" and T for "tails". The fact that the coin is fair means that the probability that heads turns up is the same as that of tails turning up, i.e.

$$P(\{H\}) = P(\{T\}) = \frac{1}{2} .$$

Rolling a die

The next example is also classical: rolling a fair die. Here the sample space is naturally given by

$$\Omega \stackrel{\text{def}}{=} \{1, 2, 3, 4, 5, 6\} .$$

The die being fair means that the probabilities of any of the numbers 1,2,3,4,5 or 6 turning up is equal to $\frac{1}{6}$, i.e.

$$P(\{1\}) = P(\{2\}) = P(\{3\}) = P(\{4\}) = P(\{5\}) = P(\{6\}) = \frac{1}{6} .$$

4.3 Elementary Combinatorial Problems

Many problems in elementary probability involve counting techniques. These techniques are studied in combinatorics and we present the most simple of them here.

4.3.1 The Basic Counting Principle

The basic problem can be described as follows. Suppose we are given a collection of finite sets $A_1, \cdots, A_r$ with respective cardinalities $n_1, \ldots, n_r$. In how many ways can we select an ordered tuple

$$(a_1, \ldots, a_r) \in A \overset{\text{def}}{=} A_1 \times \cdots \times A_r \ .$$

This is of course the same as asking what is the cardinality of A. The *Basic Counting Principle* reads:

Lemma 4.2 *The cardinality of A is given by*

$$\#A = n_1 \cdot \ldots \cdot n_r \ . \tag{4.1}$$

Proof For $r = 1$ the statement is clear. For $r \geq 2$ we proceed by induction and assume the statement is true for $r - 1$, i.e.

$$\#A_1 \times \cdots \times A_{r-1} = n_1 \cdot \ldots \cdot n_{r-1} \ . \tag{4.2}$$

We distinguish two cases. First, $\#A_r = 1$, i.e. $A_r = \{a\}$ for some a. In this case it is clear that

$$(a_1, \ldots, a_{r-1}) \mapsto (a_1, \ldots, a_{r-1}, a) \tag{4.3}$$

defines a bijection[4] between $A_1 \times \cdots \times A_{r-1}$ and $A = A_1 \times \cdots \times A_{r-1} \times A_r$. Since $n_r = 1$ and (4.2) holds we obtain (4.1).
If A_r is given by

$$A_r = \{a_{r,1}, \ldots, a_{r,n_r}\}$$

then we can write A as the disjoint union of n_r sets

$$A_1 \times \cdots \times A_{r-1} \times \{a_{r,1}\}, \cdots, A_1 \times \cdots \times A_{r-1} \times \{a_{r,n_r}\} \ ,$$

each of which has cardinality $n_1 \cdot \ldots \cdot n_{r-1}$ by the first step. It follows that this union has cardinality $n_1 \cdot \ldots \cdot n_{r-1} \cdot n_r$, proving the claim. □

---

[4]Let $f : A \to B$ be a function from set A to set B. Recall that f is an *injection* if for $x \neq y$ we have $f(x) \neq f(y)$. It is a *surjection* if for each $z \in B$ we find an $x \in A$ such that $f(x) = z$. Finally, it is a *bijection* if it is both an injection and a surjection. By definition if f is a bijection, then A and B have the same cardinality.

4.3.2   Urn Models

In the following assume we are given an urn containing n different balls, represented by a set A. The type of problem we look at now involves counting the number of ways in which one can subsequently draw, or *sample*, r balls from A. There are two features which need to be specified before the problem is well defined:

- *with or without replacement*: we need to specify whether or not after each draw we replace the ball in the urn, so that the same ball could be drawn several times; and

- *order relevance or irrelevance*: we need to make clear whether or not the order in which we draw the different balls matters. The situation where order does not matter is usually associated with the case when the balls in the urn are indistinguishable.

Sampling with replacement (order matters)

The simplest case is where we sample with replacement in such a way that order matters. A typical element is then an ordered r-tuple $(a_1, \ldots, a_r) \in = A \times \cdots \times A$. The following result is an immediate consequence of the Basic Counting Principle.

Proposition 4.3 *There are n^r ways to subsequently draw r balls from an urn containing n balls if the order in which the balls are drawn matters and after each draw we replace the chosen ball in the urn.*

Sampling without replacement (order matters)

The next simplest case is where we sample without replacement and keep track of the order in which the balls were drawn. Before stating the result, we need a definition. For each $n \in \{1, 2, \ldots\}$ the *n-factorial* is defined as

$$n! \stackrel{\text{def}}{=} n \cdot (n-1) \cdot (n-2) \cdots 2 \cdot 1.$$

Furthermore, set $0! \stackrel{\text{def}}{=} 1$.

Proposition 4.4 *There are*

$$\frac{n!}{(n-r)!} = n(n-1) \ldots (n-r+1)$$

ways to subsequently draw r balls from an urn containing n balls if the order in which the balls are drawn matters and after each draw we do not replace the chosen ball in the urn.

Proof The first ball is drawn from a set with n elements. Since the ball is not replaced the second ball is drawn from a set with $n - 1$ elements. In this way the i-th ball is drawn from a set with $n - i + 1$ elements until the r-th ball is drawn from a set with $n - r + 1$ elements. According to the basic counting principle, this can be done in

$$n(n - 1) \ldots (n - r + 1)$$

ways. □

A special case arises when we look at the case $r = n$. In this case $\frac{n!}{(n-r)!} = n!$. We can view $n!$ as the number of possible *permutations* or *ordered arrangements* of the n balls.

Corollary 4.5 *The number of permutations of a set with n elements is $n!$.*

Sampling without replacement (order does not matter)

Consider the drawing of r balls from our urn without replacement and such that order does not matter. As is easily seen this corresponds to the selection of a subset of A with r elements. Hence, counting the number of ways in which we can draw r balls without replacement and in such a way that order does not matter is the same as counting the number of subsets of A containing r elements.

Proposition 4.6 *There are*

$$\binom{n}{r} \stackrel{def}{=} \frac{n!}{(n - r)!r!}$$

ways to subsequently draw r balls from an urn containing n balls if the order in which the balls are drawn does not matter and after each draw we do not replace the chosen ball in the urn.

Proof We already know that we can draw r balls without replacement such that order matters in $\frac{n!}{(n-r)!}$ different ways. Assume $(a_1, \cdots, a_r)$ is such an ordered sample. Then, if order does not matter, we will identify each of the $r!$ possible permutations of this ordered sample with this particular selection. It follows that, if order does not matter, to obtain the number of possible choices we need to divide $\frac{n!}{(n-r)!}$ by $r!$, which is the assertion. □

Corollary 4.7 *The number of subsets of A containing r elements is equal to*

$$\binom{n}{r} \stackrel{def}{=} \frac{n!}{(n - r)!r!} \ .$$

It is easy to see that the following relationship holds.

$$\binom{n+1}{j} = \binom{n}{j} + \binom{n}{j-1}. \tag{4.4}$$

Because of the following result, known as the *binomial formula*, the numbers $\binom{n}{k}$ are called *binomial coefficients*.

Lemma 4.8 *For any numbers $x, y \in \mathbb{R}$ and $n \in \mathbb{N}$ we have*

$$(x+y)^n = \sum_{j=0}^{n} \binom{n}{j} x^{n-j} y^j .$$

Proof For $n = 1$ the assertion is trivially verified. For $n \geq 1$ we proceed by induction. Hence, assume the binomial formula is valid for n. Observe now that doing some re-indexing and using (4.4) we get

$$
\begin{aligned}
(x+y)^{n+1} &= (x+y)(x+y)^n \\
&= (x+y) \sum_{j=0}^{n} \binom{n}{j} x^{n-j} y^j \\
&= \sum_{j=0}^{n} \binom{n}{j} x^{n+1-j} y^j + \sum_{j=0}^{n} \binom{n}{j} x^{n-j} y^{j+1} \\
&= \sum_{j=0}^{n} \binom{n}{j} x^{n+1-j} y^j + \sum_{j=1}^{n+1} \binom{n}{j-1} x^{n+1-j} y^j \\
&= x^{n+1} + \sum_{j=0}^{n} [\binom{n}{j} + \binom{n}{j-1}] x^{n-j+1} y^j + y^{n+1} \\
&= x^{n+1} + \sum_{j=0}^{n} \binom{n+1}{j} x^{n-j+1} y^j + y^{n+1} \\
&= \sum_{j=0}^{n+1} \binom{n+1}{j} x^{n+1-j} y^j .
\end{aligned}
$$

This proves the assertion for $n + 1$, completing the proof. $\square$

As a consequence we obtain a formula for the cardinality of the power set, i.e. the set of all subsets, of a finite set.

Corollary 4.9 *Let A be a set of cardinality n. Then the power set 2^A of A has cardinality 2^n.*

Proof The result follows from Corollary 4.7 and the following application of the binomial formula:

$$2^n = (1+1)^n = \sum_{j=0}^{n} \binom{n}{j} 1^{n-j} 1^j = \sum_{j=0}^{n} \binom{n}{j} = \#2^A .$$

□

Sampling with replacement (order does not matter)

Surprisingly enough this is the most difficult of the four drawing problems described above.

Proposition 4.10 *There are*

$$\binom{n-1+1}{r}$$

ways to subsequently draw r balls from an urn containing n balls if the order in which the balls are drawn does not matter and after each draw we replace the chosen ball in the urn.

Proof For ease of notation we assume that

$$A = \mathbb{N}_n \overset{\text{def}}{=} \{1, 2, \ldots, n\} .$$

The outcome of drawing r balls from an urn, each time replacing the drawn ball and not caring about the order can easily be seen to be representable by an ordered r-tuple $(a_1, \ldots, a_r) \in \mathbb{N}_n$ such that $a_1 \leq a_2 \leq \ldots \leq a_r$. This means that the set of all possible outcomes can be described by the set

$$B \overset{\text{def}}{=} \{(b_1, \ldots, b_r) \in \mathbb{N}_n ; b_1 \leq b_2 \leq \ldots \leq b_r\} .$$

Define the set

$$C \overset{\text{def}}{=} \{(c_1, \ldots, c_r) \in \mathbb{N}_{n+r-1} ; c_1 < c_2 < \ldots < c_r\} ;$$

and set

$$f(b_1, \ldots, b_r) \overset{\text{def}}{=} (b_1, b_2 + 1, \ldots, b_i + i - 1, \ldots, b_r + r - 1) .$$

Then, it is easily seen that f defines a bijection between B and C. Therefore, B and C have the same cardinality. But C can be interpreted as the set of all outcomes of the experiment of drawing r balls of an urn with $n + r - 1$ balls replacing each of the drawn balls and where order does not matter. The cardinality of this set was seen to be $\binom{n+r-1}{r}$ in Proposition 4.10. □

Order in which colors are drawn does not matter

Denote the set of all balls by U. The random experiment consists of choosing n balls from the urn. This is equivalent to selecting a subset of size n of the set U of all balls. Thus, the sample space would naturally be the set of all subsets of B of size n, i.e.

$$\Omega = \{A; A \subset U \text{ and } \#A = n\}.$$

By the results of the previous section

$$\#\Omega = \binom{N}{n}.$$

All of the subsets A of U of size n are equally likely to be chosen, i.e.

$$P(A) = \frac{1}{\binom{N}{n}}.$$

What is the probability that we choose a subset $A \in \Omega$ containing k red balls? Such a subset contains k red balls and $n - k$ black balls. We can choose $\binom{R}{k}$ different subsets of size k of red balls and $\binom{N-R}{n-k}$ different subset of size $n - k$ of black balls. Therefore, there are $\binom{R}{k} \cdot \binom{N-R}{n-k}$ subsets of size n containing exactly k red balls. We have thus proved the following result.

Proposition 4.12 *Consider the event $E \stackrel{\text{def}}{=} \{A \in \Omega; A \text{ contains } k \text{ red balls}\}$. We then have:*

$$P(E) = \frac{\binom{R}{k} \cdot \binom{N-R}{n-k}}{\binom{N}{n}}.$$

Order in which colors are drawn matters

We will now use formula (4.5) in order to investigate the following problem. Assume that after each draw we record whether the ball drawn was red or black. After n draws we can represent our record by an n-tuple

$$\omega = (\omega_1, \ldots, \omega_n) \in \Omega \stackrel{\text{def}}{=} \{r, b\}^n,$$

where $\omega_i = r$ means that the i-th ball was red and $\omega_i = r$ that it was black. We would like to determine the probability for each of the possible outcomes of this new experiment. To that effect set for any $\omega \in \Omega$,

$$A_i(\omega) \stackrel{\text{def}}{=} \{\nu \in \Omega; \nu_i = \omega_i\},$$

A_i consists of all records in which the i-th entry coincides with the i-th entry of ω. Obviously,

$$\{\omega\} = \cap_{i=1}^n A_i(\omega).$$

Setting $A_i \stackrel{\text{def}}{=} A_i(\omega)$ we know from (4.5) that

$$P(\omega) = P(A_1)P(A_2|A_1)P(A_3|A_1 \cap A_2) \cdots P(A_n|\cap_{i=1}^{n-1} A_i) \,. \qquad (4.6)$$

Proposition 4.13 *For any $\omega \in \Omega$ we have*

$$P(\omega) = \frac{\prod_{i=0}^{r(\omega)-1}(R-l) \prod_{j=0}^{b(\omega)-1}(B-j)}{\prod_{l=0}^{n-1}(N-l)} \,, \qquad (4.7)$$

where $r(\omega)$ denotes the total number of drawn red balls and $b(\omega) \stackrel{\text{def}}{=} n - r(\omega)$ the total number of drawn black balls.
In particular, let A be any event in Ω and $\tau : \{1,\ldots,n\} \to \{1,\ldots,n\}$ a permutation. For any $\omega \in \Omega$ set

$$\omega^\tau \stackrel{\text{def}}{=} (\omega_{\tau(1)}, \ldots, \omega_{\tau(n)}) \,,$$

and

$$A^\tau \stackrel{\text{def}}{=} \{\omega^\tau; \omega \in A\} \,,$$

Then,

$$P(A^\tau) = P(A)$$

for all $\omega \in \Omega$.

Proof To simplify notation set $r \stackrel{\text{def}}{=} r(\omega)$, $b \stackrel{\text{def}}{=} b(\omega) = n - k$. Start by assuming that all red balls were drawn in the first k draws. We will write equation (4.6) as

$$P(\omega) = [P(A_1) \cdots P(A_r|\cap_{i=1}^{r-1} A_i)] \cdot [P(A_{r+1}|\cap_{i=1}^{r} A_i) \cdots P(A_n|\cap_{i=1}^{n-1} A_i)] \,, \quad (4.8)$$

and evaluate the terms in the square brackets separately. Since at the start there are N balls, R of which are red, the probability of first drawing a red ball is obviously

$$P(A_1) = \frac{R}{N} \,.$$

For all $l < r$ we know that $P(A_{l+1}|\cap_{i=1}^{l} A_i)$ represents the probability of drawing a red ball in the $l+1$-th draw knowing that all the previously drawn balls were also red. Thus, the total number of balls to draw from is $N - l$ and the number of red balls available for drawing is $R - l$. It follows that

$$P(A_{l+1}|\cap_{i=1}^{l} A_i) = \frac{R-l}{N-l} \,.$$

Hence, evaluation of the first square bracket is easily seen to yield

$$[P(A_1) \cdots P(A_r|\cap_{i=1}^{r-1} A_i)] = \frac{\prod_{i=0}^{r-1}(R-i)}{\prod_{l=0}^{r-1}(N-l)} \,. \qquad (4.9)$$

We proceed to evaluate the second square bracket. Knowing that in the first r draws only red balls were drawn, the probability of drawing a black ball on the $r+1$-th draw will be

$$P(A_{r+1}| \cap_{i=1}^{r} A_i) = \frac{B}{N-r} \, ,$$

since the total of balls will now be $N-r$ and the number of black balls will still be B.

For each $j < b$ we know that $P(A_{r+j+1}| \cap_{i=1}^{r+j} A_i)$ represents the probability of drawing a black ball knowing that previously r red balls and l black balls have been drawn. The total number of balls in this situation is $N-r-l$ and the total number of black balls is $B-l$. Therefore, we obtain

$$P(A_{r+j+1}| \cap_{i=1}^{r+j} A_i) = \frac{B-l}{N-r-l} \, .$$

It follows that evaluation of the second square brackets yields

$$[P(A_{r+1}| \cap_{i=1}^{r} A_i) \cdots P(A_n| \cap_{i=1}^{n-1} A_i)] = \frac{\prod_{j=0}^{b-1}(B-j)}{\prod_{l=0}^{r-1}(N-r-l)} \, . \tag{4.10}$$

Inserting expressions (4.9) and (4.10) in equation (4.8) yields formula (4.7) or this special ω in which the first k draws are red and the subsequent $n-k$ draws are black.

It is now easy to convince oneself that a permutation of the $(\omega_1, \dots, \omega_n)$ will not alter the result. Try for instance the above special case just switching the r-th and $r+1$-th entry. Since we can get any ω as a permutation of an ω in the special form above, the proposition is proved.   □

Corollary 4.14 *Setting*

$$C_i \overset{def}{=} \{\omega \in \Omega; \omega_i = r\}$$

we have

$$P(C_i) = \frac{R}{N} \, ,$$

for all i. Moreover, for all $i \neq j$ we have

$$P(C_i \cap C_j) = \frac{R}{N} \frac{R-1}{N-1} \, .$$

Proof Let $\tau : \{1, \dots, n\} \rightarrow \{1, \dots, n\}$ be any permutation which interchanges i and j. Thus, $\tau(i) = j$ and $\tau(j) = i$. Then, as is easily seen

$$C_i = C_j^\tau \, .$$

By the above proposition we have for all i and j

$$P(C_i) = P(C_j) \, .$$

Since as argued in the proof of the above proposition, $P(C_1) = \frac{R}{N}$ holds we obtain the first assertion.

Take any permutation $\tau : \{1, \ldots, n\} \to \{1, \ldots, n\}$ which sends i to 1 and j to 2. Then

$$(C_i \cap C_j)^\tau = C_i^\tau \cap C_j^\tau = C_1 \cap C_2 \,.$$

It follows that

$$P(C_i \cap C_j) = P(C_1 \cap C_2) \,.$$

We now evaluate the expression on the right. Since by (4.5) we have

$$P(C_1 \cap C_2) = P(C_1)P(C_2|C_1)$$

we obtain from the reasoning in the proof of the above proposition that

$$P(C_1 \cap C_2) = \frac{R}{N} \frac{R-1}{N-1} \,.$$

$\square$

Concluding Remarks and Suggestions for Further Reading

In this chapter we have described the first concepts of finite probability spaces. In Chapter 13 we will give a brief introduction to general probability spaces appropriate for our objectives. The serious student of mathematical finance, however, cannot avoid studying the measure theoretic approach to probability in more detail. Very readable introductions are provided in [3], [36] or [55]. A useful general reference and something of a classic is [8].

Chapter 5

Random Variables

There is a special department of hell for students of probability. In this department there are many typewriters and many monkeys. Every time that a monkey walks on a typewriter, it types by chance one of Shakespeare's sonnets.

B.A. Russell

This chapter is devoted to the study of random variables. These are the objects which will permit us to model economic variables — such as securities prices — in our stochastic economy. We will find it convenient for later applications to stress the vector space structure on the set of random variables.

5.1   Random Variables and their Distributions

Let $\Omega = \{\omega_1, \omega_2, \ldots, \omega_n\}$ be a finite sample space. Any function $X : \Omega \to \mathbb{R}$ is called a *random variable* or more precisely a *real-* or *scalar-valued random variable*. The reason for giving such a function a special name is to stress the fact that we are dealing with a random experiment and that the values X takes depend on the outcome of that experiment[1].

There are situations where the true state of the experiment will only reveal itself through one or several random variables. We will be able to observe or perceive the true outcome of the experiment only indirectly through these magnitudes. Random variables are therefore sometimes called *random signals* or *random outputs*.

---

[1] In contrast to finite probability spaces as have been defined here, in general probability spaces not every function will qualify as a random variable. We will consider more general probability spaces in Chapter 13.

5.1.1   Indicator Functions

An important example of a random variable is the *indicator function* 1_A corresponding to an event A. It is defined by

$$1_A(\omega) \stackrel{\text{def}}{=} \left\{ \begin{array}{ll} 1 & \text{if } \omega \in A \\ 0 & \text{if } \omega \in A^c. \end{array} \right.$$

This random variable detects to which of the sets A or A^c each element ω belongs to by assigning it the value 1 or 0, respectively. The constant random variable assigning to each elementary event the value 1 is of course the indicator function of the event Ω. We sometimes write 1 instead of the more cumbersome 1_Ω.

The following result collects some useful facts about indicator functions. The easy proof is left to the reader.

Lemma 5.1 *Let A and B be subsets of Ω. Then*

a) $1_{A^c} = 1 - 1_A;$

b) $1_{A \cap B} = 1_A 1_B;$

c) $1_A^2 = 1_A;$

d) $1_{A \cup B} = 1_A + 1_B - 1_A 1_B;$

e) *If $A \cap B = \emptyset$, then $1_{A \cup B} = 1_A + 1_B;$*

f) *If $A \subset B$ holds, then $1_A \leq 1_B$.*

5.1.2   Stochastic Processes

In finance we often consider random quantities which change with the passage of time. Such quantities are called stochastic processes. Let a subset I of $[0, \infty)$ be given. In this book, we will always interpret I as a set of points in time. A *stochastic* or *random process* indexed by I is a collection $(X_t)_{t \in I}$ of random variables, i.e. for each $t \in I$ a random variable $X_t : \Omega \to \mathbb{R}$ is given. If no confusion seems possible we will always use the shorter notation (X_t).

If I is an interval, (X_t) is said to be a continuous -time stochastic process. If I is discrete (finite) (X_t) is termed a discrete (finite)-time stochastic process. In this book we will consider only finite-time stochastic processes. We will study stochastic processes in more detail when we deal with multi-period models later on. For the time being we will just need the terminology.

5.1.3 Cumulative Distribution of a Random Variable

A random variable $X : \Omega \to \mathbb{R}$ is a function which depends on the outcome ω of a random experiment. It is therefore of utmost interest to use the knowledge we have of the probability laws governing the random experiment to obtain information about the likelihood of that random variable taking a particular value or range of values. Hence, given any subset $A \subset \mathbb{R}$ we ask: what is the probability that the value of X turns out to belong to the set A? Obviously, the answer is

$$P(X \in A) \overset{\text{def}}{=} P(X^{-1}(A)) .$$

We will frequently write expressions like $P(X \le x)$ or $P(X \ne x)$ instead of the correct but more cumbersome $P(\{\omega \in \Omega; X(\omega) \le x\})$ and $P(\{\omega \in \Omega; X(\omega) \ne x\})$, respectively.

Given a random variable $X : \Omega \to \mathbb{R}$ we define its *cumulative distribution function* $F_X : \mathbb{R} \to [0, 1]$ by setting

$$F_X(x) \overset{\text{def}}{=} P(X \le x) .$$

Note that

$$P(a < X \le b) = F_X(b) - F_X(a) .$$

Bernoulli or binomial distribution

Consider a probability space (Ω, P) and let an event A be given. Set $p \overset{\text{def}}{=} P(A)$. Take any $B \subset \mathbb{R}$, then

$$P(1_A \in B) = \begin{cases} p & \text{if } 1 \in B \\ 1 - p & \text{if } 1 \notin B. \end{cases}$$

We say that 1_A is Bernoulli (or: binomially) distributed with parameter p. More generally, if a random variable X has the distribution given by $P(X = 1) = p$ and $P(X = 0) = 1 - p$ for some $0 \le p \le 1$, we say that X has the *Bernoulli distribution* with parameter 1.

The cumulative distribution function for a Bernoulli distributed random variable X is given by

$$F_X(x) = P(X \le x) = \begin{cases} 1 & \text{if } x \ge 1, \\ 1 - p & \text{if } 0 \le x < 1, \\ 0 & \text{if } x < 0. \end{cases}$$

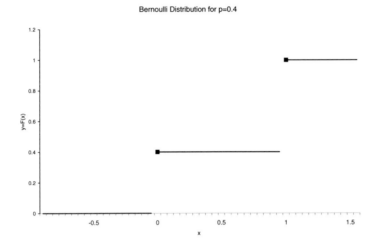

Figure 5.1: Distribution Function $F_X(x)$ of a Bernoulli Random variable for $p=0.4$.

Hypergeometric distribution

Consider the urn model of section 4.5 in which the order in which the colors were drawn did not matter. Define the random variable $X : \Omega \to \mathbb{R}$ by

$$X(A) \stackrel{\text{def}}{=} \text{number of black balls contained in } A .$$

Then by the proposition in that section

$$P(X = x) = h(x; n, R, N) \stackrel{\text{def}}{=} \frac{\binom{R}{x} \cdot \binom{N-R}{n-x}}{\binom{N}{n}}$$

if $x \in \{0, 1, \ldots, R\}$ and otherwise $P(X = x) \stackrel{\text{def}}{=} 0$. Hence, we have

$$P(X \in A) = \sum_{x \in A} h(x; n, R, N) .$$

Note that unless $x \in A \cap \{0, 1, \ldots, R\}$ the terms of the above sum are equal to zero.

Any random variable X having this distribution is called *hypergeometrically distributed* with parameters N, R and n. The cumulative distribution function is given by

$$F_X(x) = \sum_{y=0}^{x} h(y; n, R, N) .$$

5.2 The Vector Space of Random Variables

Since random variables are real-valued we may define the same sort of operations as we know for real numbers. If X and Y are (real-valued) random variables on Ω and α is any scalar number we define:

- The *sum* $X + Y$ of X and Y by $(X + Y)(\omega) = X(\omega) + Y(\omega)$,

- the *product* $X \cdot Y$ of X and Y by $(X \cdot Y)(\omega) = X(\omega)Y(\omega)$, and

- the *product* αX of α and X by $(\alpha X)(\omega) = \alpha X(\omega)$.

Thus, all operations are defined by first evaluating the involved random variables at an outcome ω and then performing the corresponding operation between real numbers. It is an easy exercise to check that, endowed with these operations, the set of random variables on Ω becomes a real vector space. We will denote it by $\mathcal{L}(\Omega)$.

The standard basis in the space of random variables

The analogue to the standard basis in $\mathbb{R}^n$ is the basis given by the set of indicator functions which detect the elementary events.

Lemma 5.2 *The vector space $\mathcal{L}(\Omega)$ has the basis*

$$\{1_{\{\omega_1\}}, 1_{\{\omega_2\}}, \ldots, 1_{\{\omega_n\}}\}.$$

Proof Note first that it is a linearly independent set. Indeed, let

$$\lambda_1 \cdot 1_{\{\omega_1\}} + \cdots \lambda_n \cdot 1_{\{\omega_n\}} = 0$$

hold. We then have for each $i \in \{1, 2, \ldots, n\}$:

$$\lambda_i = \lambda_1 \cdot 1_{\{\omega_1\}}(\omega_i) + \cdots \lambda_n \cdot 1_{\{\omega_n\}}(\omega_i) = 0,$$

proving the linear independence.
We now claim that it spans $\mathcal{V}(\Omega)$. To see this take any random variable $X : \Omega \to \mathbb{R}$. We then have

$$X = X(\omega_1) \cdot 1_{\{\omega_1\}} + \cdots + X(\omega_n) \cdot 1_{\{\omega_n\}},$$

which proves the claim. $\qquad\square$

It follows from this lemma that $\mathcal{L}(\Omega)$ is an n-dimensional vector space.

Remark 5.3 *Is easy to see that the mapping $U : \mathcal{L}(\Omega) \to \mathbb{R}^n$ given by*

$$U(X) \overset{def}{=} (X(\omega_1)), \ldots, X(\omega_n))$$

is a linear isomorphism, i.e. a linear map which is injective and surjective. By virtue of this mapping we can always identify a random variable X with the vector $U(X) = (X(\omega_1), \ldots, X(\omega_n))$. This map allows us to translate many properties of $\mathbb{R}^n$ into properties of $L(\Omega)$.

The standard representation of a random variable

Let $X : \Omega \to \mathbb{R}$ be a random variable. Because Ω has n elements, X can attain only a finite number of distinct values. Assume that the number of values is r and denote them by $x_1, \cdots, x_r$. Set furthermore for each $i \in \{1, 2, \ldots, r\}$

$$A_i \overset{\text{def}}{=} X^{-1}(x_i) .$$

We may then write:

$$X = x_1 \cdot 1_{A_1} + \cdots + x_r \cdot 1_{A_r} .$$

We will call this the *standard representation* of X. Note that if $x_j = 0$ for some j we will have $X = \sum_{i \neq j} x_i 1_A i$.

5.2.1   Linear Functionals on $L(\Omega)$

Recall that a linear subspace $\mathcal{M}$ of $L(\Omega)$ is a subset of $L(\Omega)$ such that for all $X, Y \in \mathcal{M}$ and $\lambda \in \mathbb{R}$ we have

$$X + \lambda Y \in \mathcal{M} .$$

When studying the valuation of financial contracts in later chapters we will deal with linear functionals π defined on a linear subspace $\mathcal{M}$, i.e. with functions $\pi : \mathcal{M} \to \mathbb{R}$ such that

$$\pi(X + \lambda Y) = \pi(X) + \lambda \pi(Y)$$

for all $X, Y \in \mathcal{M}$ and $\lambda \in \mathbb{R}$.

Remark 5.4 *We use the isomorphism $U : L(\Omega) \to \mathbb{R}^n$ introduced in Remark 5.3. Like any isomorphism U gives us a one-to-one correspondence between linear subspaces of $L(\Omega)$ and linear subspaces of $\mathbb{R}^n$. It also gives us a one-to-one correspondence between linear functionals defined on subspaces of $L(\Omega)$ and those defined on subspaces of $\mathbb{R}^n$.*
If $\mathcal{M}$ is a linear subspace of $L(\Omega)$ and $\pi : \mathcal{M} \to \mathbb{R}$ a linear functional, then $\mathbf{M} \overset{\text{def}}{=} U(\mathcal{M})$ is a linear subspace of $\mathbb{R}^n$ and $\phi : \mathbf{M} \to \mathbb{R}$ defined by $\phi(\mathbf{x}) \overset{\text{def}}{=} \pi(U^{-1}(\mathbf{x}))$ is a linear functional. On the other hand, if $\mathbf{M}$ is a linear subspace of $\mathbb{R}^n$ and $\phi : \mathbf{M} \to \mathbb{R}$ a linear functional, then $\mathcal{M} \overset{\text{def}}{=} U^{-1}(\mathbf{M})$ is a linear subspace of $L(\Omega)$ and $\pi : L(\Omega) \to \mathbb{R}$ defined by $\pi(X) \overset{\text{def}}{=} \phi(U(X))$ is a linear functional.

Either using the above remark and Lemma 3.1, or proceeding directly, we then have the following representation result.

Lemma 5.5 *Let $\pi : L(\Omega) \to \mathbb{R}$ be a linear functional. Then, there exists a vector $(\alpha_1, \ldots, \alpha_n) \in \mathbb{R}^n$ such that*

$$\pi(X) = \alpha_1 X(\omega_1) + \ldots + \alpha_n X(\omega_n)$$

holds for all $X \in L(\Omega)$.

5.3 Positivity on $L(\Omega)$

Positivity will play an important role throughout the whole book. The valuation of financial contacts will require us to deal with positive linear functionals on $L(\Omega)$. Here, we translate the results of Chapter 3 to the context of the vector space $L(\Omega)$

5.3.1 Positive Random Variables

We say that $X : \Omega \to \mathbb{R}$ is *positive* if

$$X(\omega_i) \geq 0$$

holds for each $i \in \{1, 2, \ldots, r\}$. X is *strictly positive* if it is positive but not identical to the zero random variable. It is *strongly positive* if

$$X(\omega) > 0$$

for all $\omega \in \Omega$. We write $X \geq 0$, $X > 0$ and $X \gg 0$ if X is positive, strictly positive and strongly positive, respectively.

Of course as in the case of $\mathbb{R}^n$ we also have an order structure given by:

- $X \geq Y$ if and only if $X - Y \geq 0$;

- $X > Y$ if and only if $X - Y > 0$;

- $X \gg Y$ if and only if $X - Y \gg 0$.

Remark 5.6 *We continue using the isomorphism $U : L(\Omega) \to \mathbb{R}^n$ of Remark 5.3. Obviously, we have that X is positive, strictly positive or strongly positive if and only if the same is true for the corresponding vector $U(X) = (X(\omega_1), \ldots, X(\omega_n))$ in $\mathbb{R}^n$.*

5.3.2 Positive Functionals

Let $\pi : \mathcal{M} \to \mathbb{R}$ be a linear functional defined on the linear subspace $\mathcal{M}$ of $L(\Omega)$. We say that π is *positive* if $\pi(X) \geq 0$ for all $X \in \mathcal{M}$ such that $X \geq 0$. If π is positive and not identical to the zero functional we call it *strictly positive*. Finally, if $\pi(X) > 0$ for all $X > 0$, we say that π is *strongly positive*.

Remark 5.7 *From Remark 5.6 we know that the isomorphism $U : L(\Omega) \to \mathbb{R}^n$, provides a one-to-one correspondence between random variables satisfying the various positivity conditions and vectors in $\mathbb{R}^n$ satisfying the corresponding conditions. Moreover, by Remark 5.4 it also provides a one-to-one correspondence between linear functionals defined on subspaces of $L(\Omega)$ and linear functionals defined on subspaces of $\mathbb{R}^n$. Not surprisingly, as the reader is invited to verify, $U : L(\Omega) \to \mathbb{R}^n$ also provides a one-to-one correspondence between (strongly), (strictly) positive functionals on subspaces of $L(\Omega)$ and (strongly), (strictly) positive functionals on subspaces of $\mathbb{R}^n$.*

From the above remark or as can be easily proved directly we have the following result on the representation of positive linear functionals on $L(\Omega)$.

Lemma 5.8 *Let $\pi : L(\Omega) \to \mathbb{R}$ be a positive linear functional. Then, there exists a positive vector $(\alpha_1, \ldots, \alpha_n) \in \mathbb{R}^n$ such that*

$$\pi(X) = \alpha_1 X(\omega_1) + \ldots + \alpha_n X(\omega_n)$$

holds for all $X \in L(\Omega)$. If the functional is strictly positive, $(\alpha_1, \ldots, \alpha_n)$ is also strictly positive. If π is strongly positive, then $(\alpha_1, \ldots, \alpha_n)$ is also strongly positive.

5.3.3   Strongly Positive Extensions

From Remark 5.7 and Corollary 3.26 we immediately obtain the following result on strongly positive extensions.

Theorem 5.9 *Let $\pi : \mathcal{M} \to \mathbb{R}$ be a strongly positive linear functional defined on the linear subspace $\mathcal{M}$ of $L(\Omega)$. Then, there exists a strongly positive extension $\tilde{\pi} : L(\Omega) \to \mathbb{R}$ to the whole space $L(\Omega)$.*
If $\mathcal{M}$ is a proper subspace of $L(\Omega)$, then there are infinitely many strongly positive extensions of π to the whole space $L(\Omega)$. Moreover, the set of strongly positive extensions is convex.

5.3.4   Optimal Extensions*

Here we just recast the results of the starred Section 3.5 into the language of random variables. The translation, which uses the isomorphism $U : L(\Omega) \to \mathbb{R}^n$ and Remark 5.7, is easy and left to the reader.
Let $\mathcal{M}$ be a proper linear subspace of $L(\Omega)$ containing a strongly positive random variable. Furthermore, let $\pi : \mathcal{M} \to \mathbb{R}$ be a strongly positive linear functional. Set

$$\mathcal{E}_\pi^+ \stackrel{\text{def}}{=} \{\psi : L(\Omega) \to \mathbb{R} \, ; \, \psi \text{ is a positive extension of } \pi\}$$

and

$$\mathcal{E}_\pi^{++} \stackrel{\text{def}}{=} \{\psi : L(\Omega) \to \mathbb{R} \, ; \, \psi \text{ is a strongly positive extension of } \pi\} \, .$$

Both these sets are not empty by Theorem 5.9. For a given random variable X we set

$$\mathcal{D}_X \{Y \in \mathcal{M} \, ; \, X \leq Y\} \, .$$

Recall that the set $\mathcal{D}_X$ is not empty since we require that $\mathcal{M}$ contains a strongly positive random variable.
From Theorem 3.31 we immediately obtain the following result.

Theorem 5.10 *Assume $\mathcal{M}$ contains a strongly positive vector. Then, for all $X \in \mathbb{R}^n$ there exist $\psi^* \in \mathcal{E}_\pi^+$ and $Y^* \in \mathcal{D}_X$ such that*

$$\psi^*(X) = \sup_{\psi \in \mathcal{E}_\pi^+} \psi(X)\,, \qquad and \qquad \pi(Y^*) = \inf_{Y \in \mathcal{D}_X} \pi(Y)\,.$$

Moreover,

$$\sup_{\psi \in \mathcal{E}_\pi^+} \psi(X) = \inf_{Y \in \mathcal{D}_X} \pi(Y)\,.$$

Remark 5.11 *As in Remark 3.33 we always have*

$$\sup_{\psi \in \mathcal{E}_\pi^{++}} \psi(X) = \sup_{\psi \in \mathcal{E}_\pi^+} \psi(X)\,.$$

5.4 Expected Value and Variance

Let X be a random variable on a probability space (Ω, P). The value of X will vary according to the outcome of the underlying random experiment. In this section we address two natural questions:

- If we repeat the experiment a large number of times what will be the "average" observed value?

- How large is the fluctuation of X around this "average" value?

5.4.1 Expected Values

Assume we repeat the random experiment several times generating each time a new value of X. What will be the average value of X? Intuitively, ω_i will occur on average p_i of the times we repeat the experiment. Hence, we would expect the average value of X to be

$$\sum_{\omega \in \Omega} X(\omega) P(\omega) = X(\omega_1) p_1 + X(\omega_2) p_2 + \ldots + X(\omega_n) p_n\,.$$

We call this average value the *expected value* of X.

The expected value of X depends on the specific assignment of probabilities to the different outcomes of the sample space. Since we will have the opportunity to consider several probability measures on a single sample space at the same time, we will often make reference to the specific probability measure in our notation for the expected value. We set

$$E_P[X] \overset{\text{def}}{=} \sum_{\omega \in \Omega} X(\omega) P(\omega)\,.$$

Sometimes the notations $\int_\Omega X(\omega) dP$ or $\int X(\omega) dP$ are used in place of $E_P[X]$.

If X is given by its standard representation

$$X = x_1 \cdot 1_{A_1} + \cdots + x_r \cdot 1_{A_r} \, ,$$

then it is easily seen that

$$
\begin{aligned}
E_P[X] &= x_1 P(X = x_1) + \ldots + x_r P(X = x_r) \\
&= x_1 P(A_1) + \ldots + x_r P(A_r) \, .
\end{aligned}
$$

The next result — which states that expectations are linear — follows easily from the definition and is left as an exercise.

Proposition 5.12 *The expectation operator is a linear functional, i.e. if P is a probability measure on the sample space Ω, X and Y are random variables, and α is a scalar, we have*

- $E_P[X + Y] = E_P[X] + E_P[Y]$, *and*

- $E_P[\alpha X] = \alpha E_P[X]$.

The expected value of a random variable X only depends on the values X assumes on the set $\{\omega \in \Omega; P(\omega) > 0\}$.

Lemma 5.13 *Set $A \stackrel{def}{=} \{\omega \in \Omega; P(\omega) > 0\}$. Then,*

$$E_P[X] = E_P[X 1_A] = \sum_{\omega \in A} X(\omega) P(\omega) \, ,$$

and

$$E_P[1_A] = 1 \, .$$

Proof Since for each $\omega \in A^c$ we have $P(\omega) = 0$ we may write

$$E_P[X] = \sum_{\omega \in \Omega} X(\omega) P(\omega) = \sum_{\omega \in \Omega} X(\omega) 1_A P(\omega) = E_P[X 1_A] \, .$$

Since also $\sum_{\omega \in \Omega} X(\omega) 1_A P(\omega) = \sum_{\omega \in A} X(\omega) P(\omega)$, the first assertion follows. The second assertion is obtained from the first as follows

$$1 = E_P[1_\Omega] = E_P[1_\Omega 1_A] = E_P[1_A] \, .$$

$\square$

Expectations are also well-behaved with respect to positivity, i.e. expectations preserve positivity.

Proposition 5.14 *For a random variable X, we have*

- $E_P[X] \geq 0$ *if* $X \geq 0$; *and*
- *if* $E_P[X] = 0$ *and* $X \geq 0$, *then* $P(X \neq 0) = 0$.

Proof The first assertion is clear from the definition. To prove the second assertion assume that X is given by its standard representation

$$X = x_1 \cdot 1_{A_1} + \cdots + x_r \cdot 1_{A_r} .$$

Assume that for some i we have $P(X = x_i) \neq 0$. By the positivity of all x_j's we get

$$E_P[X] = x_1 P(X = x_1) + \ldots + x_r P(X = x_r) \geq x_i P(X = x_i) > 0 .$$

Hence, if X is positive and $E_P[X] = 0$, we must have $P(X = x_i) = 0$ for all i. $\quad\square$

An easy and useful consequence of the positivity preserving property of expected values is the following.

Lemma 5.15 *Assume $X_1, \ldots, X_m$ are random variables, then*

$$\max\{E_P[X_1], \ldots, E_P[X_m]\} \leq E_P[\max\{X, \ldots, X_m\}]$$

Proof Observe that for any $1 \leq k \leq m$ we have

$$X_k \leq \max\{X, \ldots, X_m\} ,$$

and therefore

$$E_P[X_k] \leq E_P[\max\{X, \ldots, X_m\}] .$$

Taking the maximum with respect to k we obtain the assertion. $\quad\square$

5.4.2 Variance and Covariance

We now turn to the question of how X fluctuates around its expected value. Because of

$$X = E_P[X] + (X - E_P[X]) ,$$

we may interpret $(X - E_P[X])$ as being the the *fluctuation* of X around $E_P[X]$. As a measure of the size of this fluctuation we introduce the *variance* $Var_P(X)$ of X by setting

$$Var_P(X) \stackrel{\text{def}}{=} E_P[(X - E_P[X])^2] .$$

Later we shall interpret variance in the context of L_2 distances. The next theorem just lists some useful properties of the variance.

Theorem 5.16 *Let X be a random variable and $a, b \in \mathbb{R}$. The following assertions hold.*

a) $Var_P(X) = E_P[X^2] - E_P[X]^2.$

b) *If X is a constant on the set $A \stackrel{\text{def}}{=} \{\omega \in \Omega; P(\omega) > 0\}$, we have $Var_P(X) = 0$.*

c) $Var_P(a \cdot X + b) = a^2 \cdot Var_P(X).$

Proof To prove the first assertion set $m \stackrel{\text{def}}{=} E_P[X]$. Then, using the linearity of expectations, we get

$$
\begin{aligned}
Var_P(X) &= E_P[(X - m)^2] = E_P[X^2 - 2mX + m^2] \\
&= E_P[X^2] - 2mE_P[X] + m^2 E_P[1_\Omega] = E_P[X^2] - 2m^2 + m^2 \\
&= E_P[X^2] - m^2 ,
\end{aligned}
$$

as desired.

To prove the second assertion we assume that c is the value of the random variable on $A \stackrel{\text{def}}{=} \{\omega \in \Omega; P(\omega) > 0\}$. Then by Lemma 5.13 we have

$$E_P[X] = E_P[X 1_A] = E_P[c 1_A] = c E_P[1_A] = c .$$

We also have

$$E_P[X^2] = E_P[X^2 1_A] = E_P[c^2 1_A] = c^2 E_P[1_A] = c^2 .$$

By our first assertion we have

$$Var_P(X) = E_P[X]^2 - E_P[X^2] = c^2 - c^2 = 0 .$$

To prove the last assertion again set $m \stackrel{\text{def}}{=} E_P[X]$ and note that $E_P[aX + b] = am + b$ and thus

$$
\begin{aligned}
Var_P(aX + b) &= E_P[(aX + b - E_P[aX + b])^2] = E_P[(aX - am)^2] \\
&= a^2 E_P[(X - m)^2] = a^2 Var_P(X) .
\end{aligned}
$$

$\square$

Sometimes as a measure of dispersion the *standard deviation* $\sigma_P(X)$ is used. It is defined as

$$\sigma_P(X) \stackrel{\text{def}}{=} \sqrt{Var_P(X)} .$$

One advantage of $\sigma_P(X)$ as a measure of dispersion is that it varies proportionally to the scaling of a variable, i.e. $\sigma_P(a \cdot X) = a \cdot \sigma_P(X)$.

Covariance

Let X and Y be two random variables on (Ω, P). Their *covariance* $Cov_P(X, Y)$ is defined as

$$Cov_P(X, Y) \overset{\text{def}}{=} E_P[(X - E_P[X]) \cdot (Y - E_P[Y])] .$$

The following result establishes the link between variance and covariance. The proof is easy and is left as an exercise.

Theorem 5.17 *Let X and Y be random variables on (Ω, P). Then the following statements hold.*

a) $Cov_P(X, X) = Var_P(X)$.

b) $Cov_P(X, Y) = Cov_P(Y, X)$.

c) $Cov_P(a \cdot X + b, c \cdot Y + d) = a \cdot c \cdot Cov_P(X, Y)$.

For a collection $X_1, \ldots, X_r$ of random variables we have

$$Var_P\left(\sum_{i=1}^{r} X_i\right) = \sum_{i=1}^{r} Var_P(X_i) + \sum_{\substack{i,j=1 \\ i \neq j}}^{r} Cov_P(X_i, X_j) .$$

Note that in contrast to taking expectations, taking variances is not linear. First of all we have that $Var_P(aX) = a^2 Var_P(X)$. On the other hand we have $Var_P(X + Y) = Var_P(X) + Var_P(Y)$ only when $Cov_P(X, Y) = 0$. Two random variables X and Y for which $Cov_P(X, Y) = 0$ are said to be *uncorrelated*. The linearity of variance for uncorrelated random variables is known as the Theorem of Bienaymé.

Proposition 5.18 *For a collection $X_1, \ldots, X_r$ of uncorrelated random variables we have*

$$Var_P\left(\sum_{i=1}^{r} X_i\right) = \sum_{i=1}^{r} Var_P(X_i) .$$

We will use the following technical result, when studying expected value and variance of a hypergeometrically distributed random variable. We give the result now as an exercise in variance and covariance.

Lemma 5.19 *Let $C_1, \ldots, C_r$ be events and let a and b be numbers, such that*

- $P(C_i) = a$ *for all i; and*

- $P(C_i \cap C_j) = b$ *whenever $i \neq j$.*

Then,

$$Var_P\left(\sum_{i=1}^{r} 1_{C_i}\right) = na - n(n-1)b - n^2 a^2.$$

Proof Note first that $1_{C_i}^2 = 1_{C_i}$ and $E[1_{C_i}] = P(C_i) = a$. Therefore,

$$Var_P(1_{C_i}) = E_P[1_{C_i}^2] - E_P[1_{C_i}]^2 = a - a^2 .$$

Furthermore, we have

$$
\begin{aligned}
Cov_P(1_{C_i}, 1_{C_j}) &= E_P[(1_{C_i} - a)(1_{C_j} - a)] = E_P[1_{C_i}1_{C_j}] - a^2 \\
&= E_P[1_{C_i \cap C_j}] - a^2 = P(A_i \cap A_j) - a^2 = b - a^2 .
\end{aligned}
$$

Using Theorem 5.17 we get

$$
\begin{aligned}
Var_P\Big(\sum_{i=1}^{n} 1_{C_i}\Big) &= \sum_{i=1}^{n} Var_P(1_{C_i}) + \sum_{\substack{i,j=1 \\ i \neq j}}^{n} Cov_P(1_{C_i}, 1_{C_j}) \\
&= \sum_{i=1}^{n} (a - a^2) + \sum_{\substack{i,j=1 \\ i \neq j}}^{n} (b - a^2) \\
&= n(a - a^2) + (n^2 - n)(b - a^2) = na - n(n-1)b - n^2 a^2 .
\end{aligned}
$$

Above we have used that the numbers of pairs (i, j) with $1 \leq i, j \leq n$ is n^2. Since the number of diagonal elements is n, we have that the number of pairs (i, j) with $i \neq j$ is $n^2 - n$. $\square$

5.4.3   Functions of a Random Variable

Let $X : \Omega \to \mathbb{R}$ be a random variable on (Ω, P). If $f : \mathbb{R} \to \mathbb{R}$ is a function, then $f(X) : \Omega \to \mathbb{R}$ is also a random variable[2]. We are interested in its expected value $E_P[f(X)]$. Obviously, we have

$$E_P[f(X)] = \sum_{\omega \in \Omega} f(X(\omega))P(\omega) .$$

If X is given in its standard form

$$X = x_1 \cdot 1_{A_1} + \cdots + x_r \cdot 1_{A_r} ,$$

we can also write

$$f(X) = f(x_1) \cdot 1_{A_1} + \cdots + f(x_r) \cdot 1_{A_r} ,$$

and thus

$$E_P[f(X)] = f(x_1)P(X = x_1) + \cdots + f(x_r) \cdot P(X = x_r) .$$

---

[2] We use the notation $f(X)$ to denote the composition $f \circ X$ of f and X.

5.5 Two Examples

We calculate expected value and variance for Bernoulli and hypergeometrically distributed random variables.

5.5.1 Bernoulli Distributed Random Variables

Let X be Bernoulli distributed with parameter p. Then

- $E_P[X] = p$, and

- $Var_P[X] = p(1-p)$

To prove the first assertion we just apply the definition of expected value.

$$\begin{aligned} E_P[X] &= X(1) \cdot P(1) + X(0) \cdot P(0) \\ &= 1 \cdot p + 0 \cdot (1-p) = p. \end{aligned}$$

To prove the second assertion we also apply the definitions.

$$\begin{aligned} E_P[(X - E[X])^2] &= (X(1) - E[X])^2 \cdot P(1) + (X(0) - E[X])^2 \cdot P(0) \\ &= (1-p)^2 \cdot p + (0-p)^2 \cdot (1-p) = p \cdot (1-p). \end{aligned}$$

5.5.2 Hypergeometrically Distributed Random Variables

Consider the urn model of section 4.5 in which the order in which the colors were drawn did not matter. The random variable $X : \Omega \to \mathbb{R}$ by

$$X(A) \stackrel{\text{def}}{=} \text{number of black balls contained in } A . \tag{5.1}$$

is hypergeometrically distributed with parameters N, R, n. Then,

- $E_P[X] = \frac{nR}{N}$, and

- $Var_P[X] = n\frac{R}{N}\frac{N-R}{N}\frac{N-n}{N-1} = n \cdot p \cdot q \frac{N-n}{N-1}$ where p and q denote the proportions of black and red balls, respectively; i.e. $p \stackrel{\text{def}}{=} \frac{R}{N}$ and $q \stackrel{\text{def}}{=} \frac{N-R}{N}$.

The calculation of expected value and variance for hypergeometrically distributed random variables is more involved than for Bernoulli random variables. We will need to establish a link between the urn model in which the order in which the colors were drawn did not matter and the model in which it did matter.

Assume now that Ω is the sample space of the model in which order does matter. Denote by C_i the event "the i-th ball drawn is red". Then by Corollary 4.14 we know

- $P(C_i) = \frac{R}{N}$ for all i; and

- $P(C_i \cap C_j) = \frac{R}{N}\frac{R-1}{N-1}$ for all $i \neq j$.

Moreover,

$$Y \overset{\text{def}}{=} \sum_{i=1}^{n} 1_{C_i}$$

counts the number of red balls which were drawn in total. In this case the effect is the same as the random variable defined by (5.1). It follows that Y is hypergeometrically distributed. Hence, determining expected value and variance of Y is equivalent to determining expected value and variance of X.

Let us first look at the expected value

$$E_P[Y] = \sum_{i=1}^{n} E_P[1_{C_i}] = \sum_{i=1}^{n} P(C_i) = \sum_{i=1}^{n} \frac{R}{N} = \frac{nR}{N} \; ,$$

as claimed above. In order to calculate the variance we apply Lemma 5.19 to Y to obtain

$$
\begin{aligned}
Var_P(Y) \;\; &= \;\; Var_P\left(\sum_{i=1}^{n} 1_{C_i}\right) \\[2mm]
&= \;\; n\frac{R}{N} - n(n-1)\frac{R}{N}\frac{R-1}{N-1} - n^2\left(\frac{R}{N}\right)^2 \\[2mm]
&= \;\; \frac{nR}{N} - \frac{nR(n-1)(R-1)}{N(N-1)} - \frac{n^2R^2}{N^2} \\[2mm]
&= \;\; \frac{nRN(N-1) + nRN(n-1)(R-1) - n^2R^2(N-1)}{N^2(N-1)} \\[2mm]
&= \;\; \frac{nR(N(N-1) + N(n-1)(R-1) - nR(N-1))}{N^2(N-1)} \\[2mm]
&= \;\; \frac{nR(N^2 - Nn - NR + nR)}{N^2(N-1)} \\[2mm]
&= \;\; \frac{nR(N-n)(N-R)}{N^2(N-1)} \\[2mm]
&= \;\; n\frac{R}{N}\frac{N-R}{N}\frac{N-n}{N-1} \; .
\end{aligned}
$$

This is exactly the formula we wanted.

5.6   The L_2-Structure on $L(\Omega)$

Let (Ω, P) be a finite probability space. It is sometimes useful to impose more structure on the vector space $L(\Omega)$ of random variables on Ω. We shall describe here an "inner product structure" which turns out to be very useful in applications and also allows a geometric interpretation of many results of probability theory. However, in order to avoid some technicalities we shall assume throughout that

$$p_i = P(\omega_i) > 0$$

for all $\omega \in \Omega$.

5.6.1   The Basic Definitions

Below we introduce the notions of $\mathcal{L}_2$-product, the $\mathcal{L}_2$-norm, and the $\mathcal{L}_2$-distance. We shall define them and obtain their characteristic properties from the corresponding properties of inner product, norm and distance in classic Euclidean space.

The $\mathcal{L}_2$-product

Let X and Y be two random variables on (Ω, P). Define their L_2-product $(X|Y)_2$ by setting

$$(X|Y)_2 \stackrel{\text{def}}{=} E_P[X \cdot Y] = \sum_{i=1}^{n} X(\omega_i)Y(\omega_i)p_i .$$

Recall that we had denoted by $L(\Omega)$ the vector space of all random variables on Ω. Now, consider the mapping $T : L(\Omega) \to \mathbb{R}^n$ defined by

$$T(X) \stackrel{\text{def}}{=} (X(\omega_1)\sqrt{p_1}, \ldots, X(\omega_n)\sqrt{p_n}) .$$

The following result will allows us to translate all the properties of the Euclidean inner product on $\mathbb{R}^n$ to the L_2- product on $L(\Omega)$. The proof is easy and left as an exercise.

Lemma 5.20 *The mapping $T : L(\Omega) \to \mathbb{R}^n$ is a linear bijection and satisfies*

$$(T(X)|T(Y))_{\mathbb{R}^n} = (X|Y)_2$$

where $(x|y)_{\mathbb{R}^n}$ denotes the standard inner product on Euclidean space, i.e.

$$(x|y)_{\mathbb{R}^n} = \sum_{i=1}^{n} x_i y_i .$$

Remark 5.21 *Note that both injectivity and surjectivity of T crucially depend on our assumption that $p_i = P(\omega_i) > 0$ for all i. In case that the set $N \stackrel{\text{def}}{=} \{\omega \in \Omega; P(\omega) = 0\}$ is not empty, we still have a linear mapping but*

$$T(X) = (X(\omega_1)\sqrt{p_1}, \ldots, X(\omega_n)\sqrt{p_n})$$

will have zero-entries wherever $p_i = 0$.

With the above lemma the following result immediately follows from the corresponding properties for Euclidean space (see Appendix A).

Proposition 5.22 *Let X, Y and Z be random variables on (Ω, P) and $a, b \in \mathbb{R}$. Then the following statements hold.*

a) $(X + a \cdot Y|Z)_2 = (X|Z)_2 + a \cdot (Y|Z)_2$;

b) $(X|Y + a \cdot Z)_2 = (X|Y)_2 + a \cdot (X|Z)_2$;

c) $(X|X)_2 > 0$ for all $X \neq 0$.

The above proposition tells u that $L(\Omega)$ equipped with the L_2-product is an inner product space.

Remark 5.23 *In case that the set $N \stackrel{def}{=} \{\omega \in \Omega; P(\omega) = 0\}$ is not empty, every statement holds with the exception of the last one which only holds in the following version: $(X|X)_2 \geq 0$ for all and $(X|X)_2 = 0$ if and only if $X(\omega) = 0$ for all $\omega \neq N$.*

Recall that for $x, y \in \mathbb{R}^n$ we have the Cauchy-Schwarz inequality

$$(x|y)_{\mathbb{R}^n} \leq \left(\sum_{i=1}^{n} x_i^2\right)\left(\sum_{i=1}^{n} y_i^2\right).$$

The corresponding statement for the L_2-product is also-called the Cauchy–Schwarz inequality and can be obtained immediately from the lemma. It reads:

Proposition 5.24 *For two random variables X an Y on (Ω, P) we have*

$$|(X|Y)_2| \leq E_P[X^2]E_P[Y^2].$$

The $\mathcal{L}_2$-norm

As in Euclidean space it is convenient to introduce a norm by defining for each random variable X,

$$||X||_2 \stackrel{def}{=} \sqrt{(X|X)_2}.$$

The number $||X||_2$ is called the L_2-norm of X. The correspondence with the Euclidean norm

$$|x| = \sqrt{\sum_{i=1}^{n} x_i^2}$$

is given again by our linear bijection T.

Lemma 5.25 *The mapping $T : L(\Omega) \to \mathbb{R}^n$ satisfies*

$$|T(X)| = ||X||_2.$$

It follows that the L_2-norm satisfies the following properties:

Proposition 5.26 *For any random variables X, Y we have*

    a) $||X||_2 \geq 0$;

    b) $||X||_2 = 0$ *if and only if $X = 0$;*

    c) $||aX||_2 = |a| \cdot ||X||_2$;

    d) $||X + Y||_2 \leq ||X||_2 + ||Y||_2$.

Note that the L_2-inner product and the L_2-norm depend decisively on the particular probability measure P. For this reason we will use the special notation

$$L_2(\Omega, P)$$

when referring to $L(\Omega)$ equipped with the L_2-inner product and the L_2-norm.

Remark 5.27 *In case that the set $N \overset{\text{def}}{=} \{\omega \in \Omega; P(\omega) = 0\}$ is not empty, every statement holds with the exception of the second one which only holds in the following version: $||X||_2 = 0$ if and only if $X(\omega) = 0$ for all $\omega \neq N$.*

The L_2-distance

For two random variables X and Y on (Ω, P) define their L_2-distance $d_{2,P}(X, Y)$ by setting

$$d_{2,P}(X, Y) \overset{\text{def}}{=} \sqrt{(X - Y | X - Y)} = E_P[(X - Y)^2] \,.$$

The following properties are immediate consequences of the properties of the L_2-norm

Proposition 5.28 *Let X, Y and Z be random variables on Ω. Then the following statements hold.*

    a) $d_{2,P}(X, Y) \geq 0$;

    b) $d_{2,P}(X, Y) = 0$ *if and only if $X = Y$ holds;*

    c) $d_{2,P}(X, Y) = d_{2,P}(Y, X)$;

    d) $d_{2,P}(X, Z) \leq d_{2,P}(X, Y) + d_{2,P}(Y, Z)$.

Thus, the L_2-distance defines a distance in the strict sense of the word.

Remark 5.29 *In case that the set $N \overset{\text{def}}{=} \{\omega \in \Omega; P(\omega) = 0\}$ is not empty, every statement holds with the exception of the second one which only holds in the following version: $d_2(X, Y) = 0$ if and only if $X(\omega) = Y(\omega)$ for all $\omega \neq N$.*

5.6.2 L_2-Orthogonality

Two random variables $X, Y \in L_2(\Omega, P)$ are said to be orthogonal, denoted by $X \perp Y$, if

$$(X|Y)_2 = E_P[XY] = 0 .$$

Thus, saying that X and Y are orthogonal is the same thing as saying they are uncorrelated. The link with orthogonality in Euclidean space is given by our mapping T.

Lemma 5.30 *The random variables X and Y are orthogonal if and only if the vectors $T(X)$ and $T(Y)$ are orthogonal in Euclidean space $\mathbb{R}^n$.*

An important geometric result is known in the Euclidean space context as the *Pythagorean* theorem.

Proposition 5.31 *If X and Y are orthogonal random variables, then*

$$||X + Y||_2^2 = ||X||_2^2 + ||Y||_2^2 .$$

Proof The statement follows from

$$
\begin{aligned}
||X + Y||_2^2 &= E_P[(X + Y)(X + Y)] = E_P[X^2] + 2E_P[XY] + E_P[Y^2] \\
&= E_P[X^2] + E_P[Y^2] = ||X||_2^2 + ||Y||_2^2 .
\end{aligned}
$$

$\square$

Let $K \subset L_2(\Omega, P)$ and define its orthogonal complement by

$$K^\perp \stackrel{\text{def}}{=} \{X \in L_2(\Omega, P); X \perp Y \text{ for all } Y \in K\} .$$

As is easily seen (and we know it from Euclidean space) $K^\perp$ is always a linear subspace of $L_2(\Omega, P)$. Another fact which we can easily translate from Euclidean space to our new context is the following result on orthogonal decompositions.

Lemma 5.32 *Let $\mathcal{M}$ be a linear subspace of $L_2(\Omega, P)$. Then, $L_2(\Omega, P)$ is the direct sum of $\mathcal{M}$ and $\mathcal{M}^\perp$. In symbols*

$$L_2(\Omega, P) = \mathcal{M} \oplus \mathcal{M}^\perp .$$

This means that each X has a unique decomposition

$$X = X_{\mathcal{M}} + X_{\mathcal{M}^\perp}$$

with $X_{\mathcal{M}} \in \mathcal{M}$ and $X_{\mathcal{M}^\perp} \in \mathcal{M}^\perp$.

The linear mapping

$$P_{\mathcal{M}} : L_2(\Omega, P) \to \mathcal{M} , \quad X \mapsto X_{\mathcal{M}}$$

is called the *orthogonal projection of $L_2(\Omega, P)$ onto $\mathcal{M}$*. The orthogonal projection of a random variable X onto $\mathcal{M}$ has the beautiful property that it minimizes the distance from X to $\mathcal{M}$.

Proposition 5.33 *Of all random variables in* $\mathcal{M}$*, the orthogonal projection of* X *onto* $\mathcal{M}$ *has the smallest distance to* X*, i.e.*

$$\|X - X_\mathcal{M}\|_2 = \inf_{Y \in \mathcal{M}} \|X - Y\|_2 \,.$$

Proof Let Y be an arbitrary random variable $\mathcal{M}$. Then,

$$X - X_\mathcal{M} \in \mathcal{M}^\perp \qquad \text{and} \qquad X_\mathcal{M} - Y \in \mathcal{M} \,.$$

thus by the Pythagorean theorem we have

$$\|X - Y\|_2 = \|(X - X_\mathcal{M}) + (X_\mathcal{M} - Y)\|_2 = \sqrt{\|X - X_\mathcal{M}\|_2^2 + \|X_\mathcal{M} - Y\|_2^2} \,,$$

which is obviously minimal if $\|X_\mathcal{M} - Y\|_2 = 0$, i.e. if $Y = X_\mathcal{M}$. $\qquad\square$

5.6.3 Expected Value and Variance Revisited

Let X be a random variable. Then, we can write

$$X = E_P[X] + (X - E_P[X]) \,. \tag{5.2}$$

Observe that we can consider $E_P[X]$ as a constant random variable. Denote by $\mathcal{M}$ the linear subspace of $L_2(\Omega, P)$ consisting of all constant random variables. Thus, $E_P[X]$ and for all $C \in \mathcal{M}$ we have

$$(C|X - E_P[X])_2 = E_P[C(X - E_P[X])] = C(E_P[X] - E_P[X]) = 0 \,,$$

so that $(X - E_P[X]) \in \mathcal{M}^\perp$. Therefore, $E_P[X]$ can be viewed as the orthogonal projection of X onto the linear subspace of constant random variables. It follows that $E_P[X]$ is the constant which has minimum L_2-distance to X.

We can also interpret equation (5.2) as consisting of a prediction $E_P[X]$ for X and an error term $X - E_P[X]$. The variance of X is nothing else than the square of the norm of the error term.

Concluding Remarks and Suggestions for Further Reading

This chapter continued the discussion of the basic concepts of finite probability theory. It also gave a translation of the extension theorems for positive linear functionals described in Chapter 3 to the vector space of random variables. This is the form we will use in Chapters 6 and 9 to prove extension theorems for pricing functionals.

6.5   Completeness

The market is said to be *complete* if every alternative is attainable, i.e. if any alternative can be replicated (or "hedged") by a suitable portfolio. This is equivalent to saying that the marketed space is equal to the whole space of alternatives, i.e.

$$\mathcal{M} = \mathcal{A} .$$

6.5.1   Arrow–Debreu Securities and Completeness

For $1 \leq j \leq n$, the j-th *Arrow–Debreu security* is defined as the contingent claim which pays 1 if the state of the world is ω_j and 0 otherwise, i.e. it is the indicator function detecting the elementary event $\{\omega_j\}$. If we denote it by E_j, then

$$E_j = 1_{\{\omega_j\}} \quad , j = 1, \ldots, n .$$

The set of Arrow–Debreu securities is a vector-space basis for the space of alternatives, i.e. for each $X \in \mathcal{A}$ there is a unique n-tuple $(\lambda_1, \ldots, \lambda_n) \in \mathbb{R}^n$ such that

$$X = \lambda_1 E_1 + \lambda_2 E_2 + \ldots + \lambda_n E_n$$

holds. In fact, $\lambda_i \stackrel{\text{def}}{=} X(\omega_i)$. From this observation the following result easily follows.

Lemma 6.12 *The following statements are equivalent:*

  a) *the market is complete;*

  b) *every contingent claim is attainable; and*

  c) *every Arrow–Debreu security is attainable.*

6.5.2   The Main Result on Completeness

By definition, the market is complete if and only if the marketed space $\mathcal{M}$ coincides with the space of alternatives $\mathcal{A}$, i.e. if any alternative can be replicated by a suitable portfolio. This is only the case if the random variables $S_1^0, S_1^1, \ldots, S_1^N$ span the space of alternatives $\mathcal{A}$. Since $\mathcal{A}$ has dimension n, it follows that for completeness to hold, there must be n securities with linearly independent payoffs. It follows that two conditions are sufficient and necessary for completeness to hold:

  • there must be enough traded securities: at least n; and

  • they must be rich enough to span the whole space of alternatives: at least n of them must be linearly independent.

If we assume there are no redundant basic securities, which forces, $n \geq N+1$, and linear independence of their payoffs, which forces $n \leq N+1$, we are left with the condition $N + 1 = n$.

We summarize the above discussion in the next proposition.

Proposition 6.13 *The market is complete if and only if $N + 1 \geq n$ and there are at least n securities with linearly independent payoffs.*
In case of non-redundancy of basic securities, the market is complete if and only if $n = N + 1$.

Put in yet another way: for a market to be complete we need at least as many linearly independent basic securities as there are possible states of the world.

A remark on completeness and buy-and-hold investment strategies

One-period models can be perceived as being extremely unrealistic. But the degree of realism depends crucially on the situation we are trying to model. In fact, one-period models can be quite realistic when modelling the situation of an investor pursuing a buy-and-hold strategy, i.e. he sets up a portfolio of basic securities at time $t = 0$ and "holds" it until liquidation at his investment horizon $t = 1$. The above result on completeness has clear implications for what we can expect from a buy-and-hold strategy. In a realistic model of prices of securities, we will have a large amount of states, i.e. n will be preferably very large in our model. On the other hand, since we have a given finite number of securities — or because the investor will want to choose among a preferably small number of elementary securities out of practical considerations — N will be small in comparison to n. Hence, in realistic one-period models we will not have completeness.

6.6 The Fundamental Theorems of Asset Pricing

With what we have developed up to now, we actually have a fairly complete theory for the pricing of European alternatives. To prove our results we have essentially used the tools of linear algebra. Since it is our goal to present the theory in such a way that the reader is well prepared to approach the continuous-time case we will turn to a topic which is not strictly necessary in discrete-time models: equivalent martingale measures. We will state and prove the two Fundamental Theorems of Asset Pricing for general finite one-period models. In Chapter 11 these theorems will be proved for a general multi-period economy.

6.6.1 Strongly Positive Extensions of the Pricing Functional

Assume the market does not admit arbitrage opportunities, so that the Law of One Price holds and the pricing functional

$$\pi_0 : \mathcal{M} \to \mathbb{R}$$

is well defined and strongly positive. Note that π_0 is defined on the subspace of marketed alternatives $\mathcal{M}$ which is a proper subspace of $\mathcal{A}$ unless the market

This shows that the discounted value process of Φ is indeed a martingale with respect to Q.

The other direction is trivial since we can always think of the j-th security as a portfolio which contains only the j-th security. □

6.6.4　Equivalent Martingale Measures and the Pricing Functional

Equivalent martingale measures are useful tools for pricing. In fact, we will see that they are just a different way of representing the pricing functional.

Proposition 6.17 *Assume that the i-th security is a numeraire and that we are given an equivalent martingale measure Q. Define the linear functional $\pi_0^Q : \mathcal{A} \to \mathbb{R}$ by setting*

$$\pi_0^Q(X) \stackrel{def}{=} S_0^i \cdot E_Q[\frac{X}{S_1^i}]. \tag{6.1}$$

Then, π_0^Q is a strongly positive extension of the pricing functional π_0.

Proof The linearity of π_0^Q immediately follows from the linearity of the expectation operator. To show the strong positivity take any non-zero contingent claim X. Hence, $X(\omega) \geq 0$ for all $\omega \in \Omega$ and (since X is non-zero) there exists a j such that $X(\omega_j) > 0$. Now,

$$\begin{aligned}
\pi_0^Q(X) &= S_0^i E_Q[\frac{X}{S_1^i}] \\
&= S_0^i \cdot [\frac{X(\omega_1)}{S_1^i} \cdot q_1 + \cdots + \frac{X(\omega_n)}{S_1^i} \cdot q_n] \\
&\geq S_0^i \cdot X(\omega_j) \cdot q_j > 0.
\end{aligned}$$

This proves the strong positivity of π_0^Q.

Finally we prove that π_0^Q is an extension of π_0. Let X be an arbitrary attainable claim with replicating portfolio Φ_X. Since Q is an equivalent martingale measure we know that the discounted value process of Φ_X is also a Q-martingale. Hence,

$$\begin{aligned}
\pi_0(X) &= V_0(\Phi_X) = S_0^i \cdot \tilde{V}_0(\Phi_X) = S_0^i \cdot E_Q[\tilde{V}_1(\Phi_X)] \\
&= S_0^i \cdot E_Q[\frac{V_1(\Phi_X)}{S_1^i}] = S_0^i \cdot E_Q[\frac{X}{S_1^i}] \\
&= \pi_0^Q(X).
\end{aligned}$$

This completes the proof of the proposition. □

6.7

The result

tion 5.3.

Consider

ever, is r

attainab

Up to no

i.e. if it

hedging

Consider

positive,

The above lemma shows that to every equivalent martingale measure Q we can associate a strictly positive extension π_0^Q of the pricing functional π_0. In fact, the converse is also true. To any positive extension of the pricing functional we can associate an equivalent martingale measure. We show this next.

Proposition 6.18 *Let $\tilde{\pi}_0 : \mathcal{A} \to \mathbb{R}$ be a strictly positive extension of π_0. Then, there exists an equivalent martingale measure Q such that*

$$\tilde{\pi}(X) = \pi^Q(X) = S_0^i \cdot E_Q[\frac{X}{S_1^i}]$$

holds for every alternative $X \in \mathcal{A}$.

Any Y ∈

X. If we

impleme:

replicatic

The ques

$Y \in \mathcal{M}$

it is natu

cost of a

By apply

Proof By Corollary 3.10 we can represent $\tilde{\pi}_0$ by a strongly positive vector $x_0 = (x_0^1, \ldots, x_0^n) \in \mathbb{R}^n$, i.e. $x_0^j > 0$ for all $j = 1, \ldots, n$ and

$$\tilde{\pi}_0(X) = (X|x_0) = X(\omega_1)x_0^1 + X(\omega_2)x_0^2 + \ldots + X(\omega_n)x_0^n \tag{6.2}$$

holds for all alternatives $X \in \mathcal{A}$. Set now

$$Q(\omega_j) = q_j \stackrel{\text{def}}{=} x_0^1 \frac{S_1^i(\omega_j)}{S_0^i}.$$

We first check that Q defines a probability measure on Ω. Indeed, we see that

Theorer

attainabl

$$
\begin{aligned}
1 &= \frac{S_0^i}{S_0^i} = \frac{\pi_0(S_1^i)}{S_0^i} \\
&= \frac{S_1^i(\omega_1)x_0^1 + S_1^i(\omega_2)x_0^2 + \ldots + S_1^i(\omega_n)x_0^n}{S_0^i} \\
&= q_1 + q_2 + \ldots + q_n .
\end{aligned}
$$

Now let X be an arbitrary alternative. We then have

Theorem

for X in

pricing fu

Theorer

$$
\begin{aligned}
\pi^Q(X) &= S_0^i \cdot E_Q[\frac{X}{S_1^i}] = S_0^i \cdot [\frac{X(\omega_1)}{S-1^i} \cdot q_1 + \cdots + \frac{X(\omega_n)}{S-1^i} \cdot q_n] \\
&= S_0^i \cdot [\frac{X(\omega_n)}{S-1^i} \cdot x_0^1 \cdot \frac{S_1^{-i}(\omega_1)}{S_0^i} + \frac{X(\omega_n)}{S_1^i} \cdot \frac{S_1^i(\omega_n)}{S_0^i}] \\
&= x_0^1 \cdot X(\omega_1) + \cdots + x_0^n \cdot X(\omega_n) \\
&= \pi_0(X) .
\end{aligned}
$$

where $\mathcal{E}_{\pi_1}^+$

of π_0, res

This concludes the proof. $\square$

6.6.

Esser
prove
of the
oppo
comp
arbitr
This
usual
The e
First

Theo
An eq
only i

The c
of Ass

Theo
Assur
with

Exist

If the
is also
of Ass
not ac
is con
Recall

where
risk-ne

Hence
can be

7.1.4   Increase of Information and Refinements

Intuitively, increasing the level of detail of information implied by $\mathcal{P}$ corresponds to going over to a partition $\mathcal{Q}$ which has more observables than $\mathcal{P}$, i.e. $\mathcal{A}(\mathcal{P}) \subset \mathcal{A}(\mathcal{Q})$. In other words, by going from $\mathcal{P}$ to $\mathcal{Q}$, we will be able to say for a larger collection of events whether they occurred or not.

Alternatively, we might say that $\mathcal{Q}$ has a greater informational content than $\mathcal{P}$ if whenever information is revealed through $\mathcal{Q}$ we will know more about the true outcome than when $\mathcal{P}$ is revealed. When the information $\mathcal{P}$ is revealed to us, we are told to which of the atoms of $\mathcal{P}$, say A, the true outcome ω belongs. If $\mathcal{Q}$ has a larger informational content, then it can only be that when learning about $\mathcal{Q}$ we know that ω belongs to a subset of A. Thus, one could argue that $\mathcal{Q}$ has a larger informational content than $\mathcal{Q}$ if each atom of $\mathcal{Q}$ is contained in an atom of $\mathcal{P}$. The following lemma shows that both intuitions are equivalent.

Lemma 7.7 *The three following statements are equivalent:*

a) $\mathcal{A}(\mathcal{P}) \subset \mathcal{A}(\mathcal{Q})$;

b) *Each atom of $\mathcal{P}$ is the union of atoms in $\mathcal{Q}$;*

c) *Each atom of $\mathcal{Q}$ is contained in an atom of $\mathcal{P}$.*

Proof The equivalence of b) and c) is trivial. To prove the equivalence of b) and a) note that each atom of $\mathcal{P}$ is a union of atoms of $\mathcal{Q}$ if and only if every atom of $\mathcal{P}$ is a $\mathcal{Q}$-observable, i.e. if and only if $\mathcal{P} \subset \mathcal{A}(\mathcal{Q})$. Since, $\mathcal{A}(\mathcal{P})$ is the smallest algebra containing $\mathcal{P}$, this is equivalent to $\mathcal{A}(\mathcal{P}) \subset \mathcal{A}(\mathcal{Q})$. □

A partition $\mathcal{Q}$ is called a *refinement* of $\mathcal{P}$ if each atom in $\mathcal{P}$ is contained in an atom of $\mathcal{Q}$, i.e. we get $\mathcal{Q}$ from $\mathcal{P}$ by breaking down the atoms in $\mathcal{P}$ into smaller units. If $\mathcal{Q}$ is a refinement of $\mathcal{P}$, we also say that $\mathcal{Q}$ is *finer* than $\mathcal{P}$, or that $\mathcal{P}$ is *coarser* than $\mathcal{Q}$. If $\mathcal{Q}$ is finer than $\mathcal{P}$ we write $\mathcal{P} \preceq \mathcal{Q}$.

Example 7.8 *We illustrate this again using the example of rolling a single die with sample space $\Omega = \{1, 2, 3, 4, 5, 6\}$. Assume the informational detail is now given by the partition $\mathcal{P} = \{\{1, 2, 3, 4\}, \{5, 6\}$, corresponding to being told whether the number was smaller than or equal to four or not. The partition $\mathcal{Q}$, defined by $\mathcal{Q} = \{\{1, 2\}, \{3, 4\}, \{5, 6\}$, corresponds to a more detailed revelation of the outcome of the experiment and therefore is a refinement of $\mathcal{P}$.*

In light of the above discussion, if we use algebras to describe information it is natural to say that an algebra $\mathcal{A}$ is a *refinement* of an algebra $\mathcal{B}$ if $\mathcal{B} \subset \mathcal{A}$, i.e. if revelation of $\mathcal{A}$ will also imply knowledge of $\mathcal{B}$.

7.2   Random Variables and Measurability

Let Ω be the sample space of some random experiment. A random variable $X :$ $\Omega \to \mathbb{R}$ can be evaluated once the random experiment is performed and we are told the exact outcome ω.

But what if we learn about the exact outcome of the experiment only through revelation of a partition $\mathcal{P} = \{A_1, A_2, \ldots, A_r\}$? Can we infer which value X has taken even though we do not know the exact true outcome? Not surprisingly the answer is that we will be able to evaluate X if and only if X is constant on the atoms of $\mathcal{P}$.

Proposition 7.9 *The random variable X can be evaluated after revelation of a partition $\mathcal{P}$ if and only if X is constant on each of the atoms of $\mathcal{P}$.*

Proof Let ω be the true outcome. It is clear that if knowledge of the atom A of $\mathcal{P}$ to which ω belongs to suffices to evaluate $X(\omega)$, we must have that X is constant on A. On the other hand, if X is constant on each of the atoms of $\mathcal{P}$ it will suffice to know to which of them ω belongs to evaluate $X(\omega)$.   □

Motivated by the above lemma, we will say that $X : \Omega \to \mathbb{R}$ is $\mathcal{P}$-*measurable* or $\mathcal{P}$-*observable* if it is constant on the atoms of $\mathcal{P}$.

Observe that, by definition, if $\mathcal{P} = \{\Omega\}$, and X is $\mathcal{P}$-measurable, then X must be a constant. Also, $\mathcal{P}(X) = \{\{\omega_1\}, \{\omega_2\}, \ldots, \{\omega_n\}\}$ means that X takes a different value on each of the elementary events $\{\omega_i\}$.

Measurability with respect to an algebra

We have already seen the important relationship between partitions and algebras. The following result gives a characterization of $\mathcal{P}$-measurability in terms of the algebra of observables $\mathcal{A}(\mathcal{P})$.

Lemma 7.10 *Let X be a random variable with*

$$X(\Omega) = \{x_1, \ldots, x_r\} .$$

Then, X is $\mathcal{P}$-measurable if and only if $X^{-1}(x_i) \in \mathcal{A}(\mathcal{P})$ for each $1 \le i \le r$.

Proof Set $B_i \overset{\text{def}}{=} X^{-1}(x_i)$. By definition X is $\mathcal{P}$-measurable if it is constant on the atoms of $\mathcal{P}$. Therefore, B_i must be the union of atoms of $\mathcal{P}$. This means, by construction of $\mathcal{A}(\mathcal{P})$ that B_i belongs to $\mathcal{A}(\mathcal{P})$.

On the other hand if $B_i \in \mathcal{A}(\mathcal{P})$ holds, then B_i is the union of atoms of $\mathcal{P}$. This implies that X must be constant on atoms of $\mathcal{P}$.   □

Remark 7.11 *As is easily seen, X is $\mathcal{P}$-measurable if and only if $X^{-1}(A) \in \mathcal{A}(\mathcal{P})$ for all subsets of $\mathbb{R}$.*

From the above discussion, if $\mathcal{A}$ is an algebra, it is natural to say that X is $\mathcal{A}$-measurable if $X^{-1}(x_i) \in \mathcal{A}$ for all i. We then have: X is $\mathcal{A}$-measurable if and only if it is $\mathcal{P}(\mathcal{A})$-measurable. Of course we also have, X is $\mathcal{P}$-measurable if and only if it is $\mathcal{A}(\mathcal{P})$-measurable.

Functions of several random variables and measurability

The following simple technical result is quite useful.

Lemma 7.12 *Let $X_1, \ldots, X_m$ be $\mathcal{P}$-measurable random variables on Ω. If*

$$f : \mathbb{R}^m \to \mathbb{R}$$

is a function for which the random variable $f(X_1, \ldots, X_m) : \Omega \to \mathbb{R}$ can be defined, then $f(X_1, \ldots, X_m)$ is also $\mathcal{P}$-measurable.

Proof Take any atom A of $\mathcal{P}$. Since $X_1, \ldots, X_m$ are all $\mathcal{P}$-measurable they are all constant on A. This immediately implies that $f(X_1, \ldots, X_m)$ is constant on A, proving that $f(X_1, \ldots, X_m)$ is $\mathcal{P}$-measurable.   □

Some particular cases will be of interest later on.

Corollary 7.13 *Let X and Y be $\mathcal{P}$-measurable random variables. Then,*

$$X \cdot Y, \quad \frac{X}{Y} \ (Y \gg 0), \quad \max\{X, Y\}, \quad and \quad \min\{X, Y\}$$

are also $\mathcal{P}$-measurable.

Proof As an example we prove that $\frac{X}{Y}$ is measurable. Choose

$$f(x, y) \stackrel{\text{def}}{=} \frac{x}{y}$$

and apply the lemma.   □

7.3   Linear Subspaces of $L(\Omega)$ and Measurability

Recall that $L(\Omega)$ denoted the vector space of random variables on Ω. It is useful to know that given a partition $\mathcal{P}$, the set consisting of all $\mathcal{P}$-measurable random variables is closed under the vector space operations.

Proposition 7.14 *The set*

$$L(\Omega, \mathcal{P}) \stackrel{\text{def}}{=} \{X : \Omega \to \mathbb{R}; X \text{ is } \mathcal{A}\text{-measurable }\}$$

is a linear subspace of $L(\Omega)$.

Proof For $\lambda \in \mathbb{R}$ define

$$f(x,y) \stackrel{\text{def}}{=} x + \lambda y \ .$$

Then, by Lemma 7.12 we know that if X and Y are $\mathcal{P}$-measurable random variables, then

$$f(X,Y) = X + \lambda Y$$

is also $\mathcal{P}$-measurable.　　　　　　　　　　　　　　　　　　　　　□

7.3.1　Standard Basis for $L(\Omega, \mathcal{P})$

Let the partition $\mathcal{P}$ of Ω be given by

$$\mathcal{P} = \{A_1, \dots, A_r\} \ .$$

Each of the random variables $1_{A_1}, \dots, 1_{A_r}$ is of course $\mathcal{P}$ measurable and therefore belongs to $L(\Omega, \mathcal{P})$. In fact these random variables form a basis for $L(\Omega, \mathcal{P})$.

Lemma 7.15 *The random variables* $1_{A_1}, \dots, 1_{A_r}$ *form a basis for* $L(\Omega, \mathcal{P})$, *i.e. every random variable* X *in* $L(\Omega, \mathcal{P})$ *has a unique representation*

$$X = \lambda_1 1_{A_1} + \dots + \lambda_r 1_{A_r}, \tag{7.1}$$

with $\lambda_1, \dots, \lambda_r \in \mathbb{R}$.

Proof Let

$$\lambda_1 1_{A_1} + \dots + \lambda_r 1_{A_r} = 0 \ .$$

For any i choose $\omega \in A_i$. Then, it follows that $0 = X(\omega_i) = \lambda_i$. Hence, the random variables $1_{A_1}, \dots, 1_{A_r}$ are linearly independent.
On the other hand, if X belongs to $L(\Omega, \mathcal{P})$ it is constant on each of the atoms $A_1, \dots, A_r$. With $\lambda_i \stackrel{\text{def}}{=} X(A_i)$ we see that (7.1) holds. It follows that $1_{A_1}, \dots, 1_{A_r}$ is a basis for $L(\Omega, \mathcal{P})$.　　　　　　　　　　　　□

7.3.2　When is a Subspace of $L(\Omega)$ of the Form $L(\Omega, \mathcal{P})$?*

The material presented here is not used elsewhere in the book.
It is interesting to ask which subspaces of $L(\Omega)$ are given as $L(\Omega, \mathcal{P})$ for a suitable partition $\mathcal{P}$. The answer is given in the next proposition.

Proposition 7.16 *A linear subspace $\mathcal{M}$ of $L(\Omega)$ is given as*

$$\mathcal{M} = L(\Omega, \mathcal{P})$$

for some partition $\mathcal{P}$ if and only if the following two conditions are satisfied:

- 1_Ω *belongs to $\mathcal{M}$.*

- $\max\{X, Y\}$ *belongs to $\mathcal{M}$ if X and Y belong to $\mathcal{M}$.*

Proof
Any space which is given as $L(\Omega, \mathcal{P})$ for some partition Ω satisfies these two conditions as can be verified using Corollary 7.13.
Let then $\mathcal{M}$ satisfy the two conditions. Define

$$\mathcal{A} \stackrel{\text{def}}{=} \{A \subset \Omega; 1_A \in \mathcal{M}\} \,.$$

Note that $c\mathcal{A}$ is not empty since Ω and $\emptyset$ belong to $\mathcal{A}$. Moreover, if $A \in \mathcal{A}$, then

$$1_{A^c} = 1_\Omega - 1_A \in \mathcal{M} \,,$$

so that $A^c \in \mathcal{A}$. If $A, B \in \mathcal{A}$, then

$$1_{A \cup B} = \max\{1_A, 1_B\} \in \mathcal{M} \,,$$

so that $A \cup B \in \mathcal{A}$. It follows that $\mathcal{A}$ is an algebra.
Let $\mathcal{P}$ be the unique partition generating $\mathcal{A}$. By what we have just proved,

$$L(\Omega, \mathcal{P}) \subset \mathcal{M} \,.$$

We now show that every $X \in \mathcal{M}$ belongs to $L(\Omega, \mathcal{P})$. For such an X in $\mathcal{M}$ let $\lambda > 0$ be large enough so that

$$\lambda X > 1$$

on $[X > 0]$. Then,

$$1_{[X>0]} = \max\{\lambda X, 1\} \in \mathcal{M} \,,$$

so that $[X > 0] \in \mathcal{A}$. Since $a1_\Omega \in \mathcal{M}$ it follows that $[X > a] = [X - a1_\Omega > 0] \in \mathcal{A}$ for any $a \in \mathbb{R}$. Since with X, also $-X$ belongs to $\mathcal{M}$, it follows that $[X < b] \in \mathcal{A}$ for any $b \in \mathbb{R}$. Note that X can assume only finitely many values $x_1, \ldots, x_l$. For any such value x_i we find $a < x_i < b$ such that

$$[X = x_i] = [X > a] \cap [x < b] \in \mathcal{A}$$

It follows by Lemma 7.10 that X is $\mathcal{P}$-measurable.   □

7.4 Random Variables and Information

Sometimes we only learn about the outcome of a random experiment indirectly through the observation of one or several random variables. How does this fit in the framework of partitions and algebras developed in the previous sections?

7.4.1 Information Revealed Through X

Let $X : \Omega \to \mathbb{R}$ be a random variable and assume that

$$X(\Omega) = \{x_1, \ldots, x_r\} .$$

For each $j = 1, 2, \ldots, r$ set

$$B_j \stackrel{\text{def}}{=} X^{-1}(x_j) .$$

Thus, the event B_j is just the set where X takes the value x_j. Observe that

$$\mathcal{P}(X) \stackrel{\text{def}}{=} \{B_1, B_2, \ldots, B_r\}$$

is a partition — the *standard or natural partition* generated by X. Set also

$$\mathcal{A}(X) \stackrel{\text{def}}{=} \mathcal{A}(\mathcal{P}(X)) .$$

The algebra $\mathcal{A}(X)$ is called the *standard or natural algebra* generated by X. Assume that we only learn about the outcome of our random experiment through the values X takes. This means that when the experiment is performed and the true outcome is some ω, we will only learn that $X(\omega)$ was observed. If $X(\omega) = x_j$ for some j this will tell us that B_j occurred. However, for us, the occurrence of any of the elements of B_j would have revealed the same information. This reasoning shows us that learning about the experiment through X is the same as having the informational detail of $\mathcal{P}(X)$ or equivalently of $\mathcal{A}(X)$.

Example 7.17 *We illustrate this again using the example of rolling a single die with sample space $\Omega = \{1, 2, 3, 4, 5, 6\}$. Consider the random variable $X : \Omega \to \mathbb{R}$ defined by*

$$X(\omega) \stackrel{def}{=} 1_{\{1,2\}} - 1_{\{5,6\}} .$$

This random variable can take the values 1, 0, and −1. The associated partition is easily seen to be

$$\mathcal{P}(X) = \{\{1, 2\}, \{3, 4\}, \{5, 6\}\} .$$

Assume that we only learn about the outcome of the experiment by being told the value of X. Now, if we are told that the value is 1 we infer that $\{1, 2\}$ occurred, if it is 0 we know that $\{3, 4\}$ occurred and finally if it is −1 we know that $\{5, 6\}$ occurred. Hence, learning about the experiment through X is like $\mathcal{P}(X)$ being revealed.

$\mathcal{P}$-measurability of X revisited

Lemma 7.18 *The value of X can be established upon revelation of $\mathcal{P}$ if and only if $\mathcal{P}$ is a refinement of $\mathcal{P}(X)$, or equivalently if $\mathcal{A}(\mathcal{P}) \subset \mathcal{A}(\mathcal{P}(X))$.*

Proof Assume $\mathcal{P}$ is revealed to us. Then we will know that the true outcome ω of the experiment belongs to a particular atom of $\mathcal{P}$, say A_j. In order to be able to evaluate $X(\omega)$ we will need to know to which of the atoms $B_1, \cdots, B_n$ of $\mathcal{P}(X)$ the true outcome ω belongs. But this is equivalent to finding a B_l such that $A_j \subset B_l$. In this way we find that every atom in $\mathcal{P}$ must be contained in an atom of $\mathcal{P}(X)$, i.e. $\mathcal{P}$ is finer than $\mathcal{P}(X)$.   □

7.4.2 $\mathcal{P}(X)$-Measurability Characterized

Assume we learn about the outcome of a random experiment through the observation of a single random variable X. We know that upon being told the value of X we can evaluate a second random variable Y if and only if Y is constant on the atoms of $\mathcal{P}(X)$. The following result gives another more intuitive characterization.

Proposition 7.19 *Let X be a random variable. Then, a random variable Y is $\mathcal{P}(X)$-measurable if and only if there exists a function $f : \mathbb{R} \to \mathbb{R}$ such that*

$$Y = f(X) \,.$$

Proof Let

$$X(\Omega) = \{x_1, \ldots, x_r\}$$

and set

$$B_i \stackrel{\text{def}}{=} X^{-1}(x_i) \quad \text{and} \quad C_j \stackrel{\text{def}}{=} Y^{-1}(y_j) \,.$$

Thus,

$$\mathcal{P}(X) = \{B_1, \ldots, B_r\}$$

is the natural partition of X.

Now, if Y is $\mathcal{P}(X)$-measurable, then it is constant on each of the $B1, \ldots, B_r$. Take any function $f : \mathbb{R} \to \mathbb{R}$ such that

$$f(x_i) = Y(B_i) \,.$$

For any $\omega \in \omega$ we find an i such that $\omega \in B_i$. But then

$$f(X)(\omega) = f(X(\omega)) = f(x_i) = Y(B_i) = Y(\omega) \,.$$

To prove the opposite direction assume that $Y = f(X)$ for some f. But then it is obvious that Y is constant on the atoms of $\mathcal{P}(X)$, i.e. Y is $\mathcal{P}(X)$-measurable.   □

7.4.3   Measurability with Respect to Several Random Variables

Assume now that we learn about the random experiment through the observation of a finite number m of random variables. Denote them by

$$X_1, \ldots, X_m : \Omega \to \mathbb{R} .$$

When the experiment is performed with true outcome ω we observe the vector

$$(X_1(\omega), \ldots, X_m(\omega)) \in \mathbb{R}^m .$$

The (finite) set of all possible vectors $\mathbf{v} = (v_1, \ldots, v_m) \in \mathbb{R}^m$ we can observe is given by

$$\mathbf{V} \stackrel{\text{def}}{=} \{\mathbf{v} \in \mathbb{R}^m ; \mathbf{v} = (X_1(\omega), \ldots, X_m(\omega)) \text{ for some } \omega \in \Omega\} .$$

Hence, observing the vector $\mathbf{v} \in \mathbf{V}$ is equivalent to knowing that the true outcome ω belongs to

$$C_{\mathbf{v}} \stackrel{\text{def}}{=} [X_1 = v_1, \cdots, X_m = v_m] = \cap_{i=1}^m [X_i = v_i] .$$

The collection of events

$$\mathcal{P}(X_1, \ldots, X_m) \stackrel{\text{def}}{=} \{C_{\mathbf{v}}; \mathbf{v} \in \mathbf{V}\}$$

is easily seen to be a partition. We conclude that learning about the experiment through the random variables $X_1, \ldots, X_m$ is equivalent to being told the partition

$$\mathcal{P}(X_1, \ldots, X_m) ,$$

which we call the *natural* partition generated by $X_1, \ldots, X_m$. Correspondingly, we call

$$\mathcal{A}(X_1, \ldots, X_m) \stackrel{\text{def}}{=} \mathcal{A}(\mathcal{P}(X_1, \ldots, X_m))$$

the *natural* algebra generated by $X_1, \ldots, X_m$.
The following lemma characterizes $\mathcal{P}(X_1, \ldots, X_m)$-measurability.

Proposition 7.20 *Let $X_1, \ldots, X_m$ and Y be random variables. Then, Y is measurable with respect to $\mathcal{P}(X_1, \ldots, X_m)$ if and only if there exists a function $f : \mathbb{R}^m \to \mathbb{R}$ such that*

$$Y = f(X_1, \ldots, X_m) .$$

Proof Assume that Y is measurable with respect to $\mathcal{P}(X_1,\ldots,X_m)$. Then, it is constant on the atoms $C_{\mathbf{v}}$, $\mathbf{v} \in \mathbf{V}$, defined above. Take any function $f : \mathbb{R}^m \to \mathbb{R}$ such that

$$f(\mathbf{v}) = Y(C_{\mathbf{v}}) \, .$$

For any $\omega \in \Omega$, we find a $\mathbf{v} \in \mathbf{V}$ such that $\omega \in C_{\mathbf{v}}$. But then

$$f(X_1,\ldots,X_m)(\omega) = f(X_1(\omega),\ldots,X_m(\omega)) = f(\mathbf{v}) = Y(C_{\mathbf{v}}) = Y(\omega) \, .$$

To prove the opposite direction note that if $Y = f(X_1,\ldots,X_m)$ for some function f then Y is automatically constant on the atoms $C_{\mathbf{v}}$ of $\mathcal{P}(X_1,\ldots,X_m)$, i.e. Y is $\mathcal{P}(X_1,\ldots,X_m)$-measurable. $\square$

7.5   Information Structures and Flow of Information

In the real world, as times passes the level of information increases and we learn more about the true state of the world. This is modelled by the concept of an information structure.

Let $t = 0,1,\ldots,T$ be a sequence of times. An *information structure* is a collection $\mathcal{I} = \{\mathcal{P}_0, \mathcal{P}_1, \ldots, \mathcal{P}_T\}$ of partitions such that the following three conditions are satisfied:

- $\mathcal{P}_0 = \{\Omega\}$,

- $\mathcal{P}_T = \{\{\omega_1\}, \{\omega_2\}, \ldots, \{\omega_n\}\}$, and

- $\mathcal{P}_0 \preceq \mathcal{P}_1 \preceq \mathcal{P}_2 \preceq \ldots \preceq \mathcal{P}_T$.

The first property means that at the beginning we do not know anything about the outcome of the experiment other than that it will occur. The second property means that at the end of the time horizon T all uncertainty will have been resolved and we will know the exact outcome of the experiment. The third property implies that as time progresses, more and more information is made available to us, thus each of the posterior partitions will be a refinement of the previous ones

7.5.1   Visualizing Information Structures

We illustrate by a simple example how to visualize information structures. Consider the probability space

$$\Omega = \{\omega_1, \omega_2, \omega_3, \omega_4\} \, .$$

We assume that we have three dates $t = 0, 1, 2$ at which information in the form of the partitions $\mathcal{P}_0, \mathcal{P}_1$ and $\mathcal{P}_2$ is received. The partitions are given by:

$$
\begin{aligned}
\mathcal{P}_0 &= \{\Omega\} \\
\mathcal{P}_1 &= \{\{\omega_1, \omega_2\}, \{\omega_3, \omega_4\}\} \\
\mathcal{P}_2 &= \{\{\omega_1\}, \{\omega_2\}, \{\omega_3\}, \{\omega_4\}\}
\end{aligned}
$$

We visualize the flow of information by using the tree below.

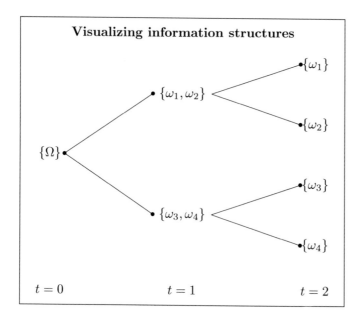

We imagine that at time $t = 0$ nothing is revealed on the outcome of the experiment. Thus, we only know that the true outcome will belong to Ω. At time $t = 1$, however, the first piece of real information arrives. Ω is divided into the two sets $\{\omega_1, \omega_2\}$ and $\{\omega_3, \omega_4\}$ and we learn to which of these two sets the outcome belongs. Assume we learned that the outcome belongs to $\{\omega_1, \omega_2\}$. At time $t = 2$ the event $\{\omega_1, \omega_2\}$ will split into the elementary events $\{\omega_1\}$ and $\{\omega_2\}$ and we will learn which of ω_1 and ω_2 is the true outcome. Assume on the other hand that at time $t = 1$ we had learned that the outcome belonged to $\{\omega_3, \omega_4\}$. Then at time $t = 2$ this event will split into the two elementary events $\{\omega_3\}$ and $\{\omega_4\}$ and we will learn whether ω_3 or ω_4 was the true outcome.

It follows that we will gradually find out about the true outcome of the experiment ω though the branch of the tree to which it belongs.

7.6   Stochastic Processes and Information Structures

Let $(X_t)_{0 \le t \le T}$ be a finite stochastic process. Assume furthermore that an information structure $\mathcal{I} = \{\mathcal{P}_0, \mathcal{P}_1, \ldots, \mathcal{P}_T\}$ is given. The stochastic process is called $\mathcal{I}$-adapted if for each t the random variable X_t is $\mathcal{P}_t$-measurable.

Interpretation 7.21 *Requiring that a process $(X_t)_{0 \le t \le T}$ be adapted to the partition $\mathcal{I}$ ensures that, when at time t the information contained in $\mathcal{P}_t$ arrives, we are able to evaluate the variable X_t.*

This is exactly the feature that we have to require if X_t were to be interpreted as the price of a security at time t: At time t the price of the security X_t at time t should be known!

7.6.1   Information Structures Given by Random Variables

Sometimes we will encounter the situation where information is revealed through a series of random variables $Y_1, Y_2, \ldots Y_T$ where Y_t becomes known at time t. In other words, at time t all we will know about the true outcome ω of the experiment will come only through the revelation of $Y_1(\omega), \ldots, Y_t(\omega)$. As discussed in Section 7.4.3, this is equivalent to knowing the partition $\mathcal{P}(Y_1, \ldots, Y_t)$.
The corresponding information structure is given by $\mathcal{I} = \{\mathcal{P}_0, \mathcal{P}_1, \ldots, \mathcal{P}_T\}$, where $\mathcal{P}_t \stackrel{\text{def}}{=} \mathcal{P}(Y_1, \ldots, Y_t)$. If (X_t) is an $\mathcal{I}$-adapted stochastic process, then we find functions $F_t : \mathbb{R}^t \to \mathbb{R}$ such that

$$X_t = F_t(X_1, \ldots, X_t) .$$

7.6.2   Predictability with Respect to an Information Structure

Assume we are given a stochastic process $(X_t)_{0 \le t \le T}$ and an information structure $\mathcal{I} = \{\mathcal{P}_0, \mathcal{P}_1, \ldots, \mathcal{P}_T\}$. We have just seen that $\mathcal{I}$-adaptedness of the random process means that $X_t(\omega)$ can be evaluated upon availability of the information contained in $\mathcal{P}_t$. We also observed that it is natural to require $\mathcal{I}$-adaptedness if the process is to describe the evolution of the price of a security.
Consider now a process $(r_t)_{0 \le t \le T}$ describing interest rates in the following sense. For $t = 0$ we just require $r_0(\omega) = 0$ for all $\omega \in \Omega$. For $t \ge 1$ the random variable r_t is to be interpreted as the rate at which money can be invested at time $t - 1$ until time t, i.e. investing 1 currency unit at time $t - 1$ gives you $(1 + r_t(\omega))$ currency units at time t, depending on what the true state of the world ω is. But this process has another peculiarity: The rate $r_t(\omega)$ is observed at time $t - 1$. Hence, the information available at that time should suffice to evaluate $r_t(\omega)$, i.e. r_t should be $\mathcal{P}_{t-1}$-measurable.

The above discussion motivates the following definition. Any random process $(X_t)_{0 \leq t \leq T}$ is said to be *$\mathcal{I}$-predictable* if

- X_0 is $\mathcal{P}_0$-measurable, and

- X_t is $\mathcal{P}_{t-1}$-measurable for $t \geq 1$.

For a predictable random process $(X_t)_{0 \leq t \leq T}$, the value of $X_t(\omega)$ is already known at time $t - 1$.

Concluding Remarks and Suggestions for Further Reading

The material in this section is important because it permits a mathematically precise description of the fact that as time elapses we learn more about the probabilistic model underlying the world. When dealing with continuous-time stochastic processes, more general concepts are needed and the technicality of the material increases considerably. In the next chapter we will deal with independence, yet another aspect of information.

References for stochastic processes are [49] or [8].

Chapter 8

Independence

> *Thus one comes to perceive, in the concept of independence, at least the first germ of the true nature of problems in probability theory.*
>
> A.N. Kolmogorov

8.1 Independence of Events

The concept of independent events formalizes a fairly natural idea. Intuitively speaking, two events A and B are said to be independent if any statement concerning the occurrence or non-occurrence of one of them does not change the probability of the other event. Independence is one of the central concepts in probability theory. In this and the next section we explain the mathematical formulation of the concept and investigate the different guises in which it appears.

8.1.1 Independent Events

Assume that a sample space $\Omega = \{\omega_1, \cdots, \omega_n\}$ and a probability measure P thereon are given. Interpreting probabilities as relative frequencies we may try to formalize the introductory remarks. Consider the following example. Suppose $P(A) = 0.4$ and $P(B) = 0.3$. In a large number of repetitions of the experiment, A will occur roughly 40% of the time. If A and B are to be independent the occurrence of A should not affect the relative frequency with which B occurs. Hence, if we examine only those trials on which A occurs, i.e. those trials where $A \cap B$ occurred, we should observe B approximately 30% of the time. It follows that $P(A \cap B) = 0.4 \cdot 0.3 = P(A)P(B)$. Similarly, the non-occurrence of A, i.e. the occurrence of A^c, will not affect the relative frequency with which B occurs. This leads to $P(A^c \cap B) = 0.6 \cdot 0.3 = P(A^c)P(B)$.

On the other hand, if $P(A \cap B) = P(A)P(B)$ and $P(A^c \cap B) = P(A^c)P(B)$, then the occurrence of A will not influence the relative frequency of B. Indeed, if we examine only those trials on which A occurs, B must occur approximately 30% of the time, in order to have $P(A \cap B) = P(A)P(B)$. Hence, the occurrence of A does not influence the relative frequency of B. Analogously, the non-occurrence of A does not influence the relative frequency of B either.

We define A and B to be *independent* if

$$P(A \cap B) = P(A)P(B)$$

holds. Observe, that this automatically implies

$$P(A^c \cap B) = P(A^c)P(B) .$$

Indeed, this follows from

$$\begin{aligned}P(A^c \cap B) &= P(B) - P(A \cap B) = P(B) - P(A)P(B) \\ &= (1 - P(A))P(B) = P(A^c)P(B).\end{aligned}$$

Example 8.1 *Assume we roll a die twice. The corresponding sample space is*

$$\Omega \overset{def}{=} \{(1,1), \dots, (1,6), \dots, (6,1), \dots, (6,6)\}$$

and every elementary event (i,j) has the same probability $P(i,j) = \frac{1}{36}$. Let A be the event "rolling the first time a 6" and B the event "rolling the second time a 2", i.e.

$$A = \{(6,1), \dots (6,6)\} \quad and \quad B = \{(1,2), \dots (6,2)\} .$$

Since $A \cap B = \{(6,2)\}$ holds, we have $P(A) = P(B) = \frac{6}{36}$ and $P(A \cap B) = \frac{1}{36}$. Therefore, $P(A \cap B) = P(A)P(B)$ and, thus, A and B are independent.

The following example may seem counter-intuitive at first.

Example 8.2 *From the definition it is immediate that any set is independent of Ω and of the empty set. In fact, you can easily verify that any set B with $P(B) = 0$ or $P(B) = 1$ is independent of any other event.*

Independent events and conditional probabilities

Let A and B be two events in ω and $P(B) > 0$. Recall that in Section 4.4 we had defined the probability of A conditional on B as

$$P(A|B) = \frac{P(A \cap B)}{P(B)} .$$

We immediately see that saying that A and B are independent, is equivalent to saying that

$$P(A|B) = P(A) .$$

Thus, if A and B are independent, knowing that B occurred does not alter our assessment of the probability for A to occur.

8.1.2   Independent Sequences of Events

Sometimes it is necessary to consider independence among more than two events. The events $A_1, \ldots, A_m$ are said to be independent if for all subsets $J \subset \{1, \ldots, m\}$ we have

$$P(\cap_{j \in J} A_j) = \prod_{j \in J} P(A_j) \, .$$

Independence of a sequence $A_1, \ldots, A_m$ corresponds to the intuition that the occurrence of any group of the events $A_1, \ldots, A_m$ should be independent of the occurrence of the remaining remaining events. Our next result gives this and other simple restatements of the definition. These restatements, however simple, should help sharpen the understanding of the concept of an independent sequence of events.

Proposition 8.3 *The following statements are equivalent:*

  a) *The events $A_1, \ldots, A_m$ are independent.*

  b) *Every finite subsequence of $A_1, \ldots, A_m$ is independent.*

  c) *Let $J_1, \ldots, J_r$ be disjoint subsets of $\{1, \ldots, m\}$ and set*

$$B_i \stackrel{\text{def}}{=} \bigcap_{j_i \in J_i} A_{j_i} \, .$$

  Then, the sequence $B_1, \ldots, B_r$ satisfies

$$P(B_1 \cap \ldots \cap B_r) = P(B_1) \cdot \ldots \cdot P(B_r) \, .$$

  d) *Let I and J be disjoint subsets of $\{1, \ldots, m\}$ and set*

$$A \stackrel{\text{def}}{=} \cap_{i \in I} A_i \qquad \text{and} \qquad B \stackrel{\text{def}}{=} \cap_{j \in J} A_j \, .$$

  Then, A and B satisfy

$$P(A \cap B) = P(A)P(B) \, .$$

Proof Statement b) is obviously just a straight reformulation of a).
To see that a) implies c) let $J_1, \ldots, J_r$ and B_i be given as in the statement of c). By the independence of $A_1, \ldots, A_m$ we have

$$
\begin{aligned}
P(B_i) &= P(\cap_{j_i \in J_i}) = \prod_{j_i \in J_i} P(A_{j_i}) \, , \\
P(A_{j_1} \cap \ldots \cap A_{j_r}) &= P(A_{j_1}) \cdot \ldots \cdot P(A_{j_r}) \, .
\end{aligned}
$$

It therefore follows, again using the independence assumption, that

$$
\begin{aligned}
P(B_1 \cap \ldots \cap B_r) &= P\left(\bigcup_{(j_1,\ldots,j_r) \in J_1 \times \ldots \times J_r} [A_{j_1} \cap \ldots \cap A_{j_r}] \right) \\
&= \prod_{(j_1,\ldots,j_r) \in J_1 \times \ldots \times J_r} P(A_{j_1} \cap \ldots \cap A_{j_r}) \\
&= \prod_{(j_1,\ldots,j_r) \in J_1 \times \ldots \times J_r} P(A_{j_1}) \cdot \ldots \cdot P(A_{j_r}) \\
&= \prod_{j_1 \in J_1} P(A_{j_1}) \cdot \ldots \cdot \prod_{j_r \in J_r} P(A_{j_r}) \\
&= P(B_1) \cdot \ldots \cdot P(B_r) \, .
\end{aligned}
$$

Statement d) is only a special case of statement c). It remains to prove that d) implies a). It is clear that for any two events A_j and A_i we have

$$
P(A_j \cap A_i) = P(A_j)P(A_i) \, .
$$

For larger subsequences we proceed by induction. Assume that for any subset $I \subset \{1,\ldots,m\}$ of r elements we have

$$
P(\cap_{i \in I}) = \prod_{i \in I} P(A_i) \, .
$$

Let J be a subset of $\{1,\ldots,m\}$ with $r+1$ elements and take any $\hat{j} \in J$. Then, $I \overset{\text{def}}{=} J \setminus \{\hat{j}\}$ has r elements. Setting $A \overset{\text{def}}{=} A_{hatj}$ and $B \overset{\text{def}}{=} \cap_{i \in I} A_i$, we see that

$$
\begin{aligned}
P(\cap_{j \in J} A_j) &= P(A \cap B) = P(A)P(B) = P(A_{\hat{j}})P(\cap_{i \in I} A_i) \\
&= P(A_{\hat{j}}) \prod_{i \in I} P(A_i) = \prod_{j \in J} P(A_j) \, .
\end{aligned}
$$

This proves that $A_1,\ldots,A_m$ are independent.   □

The following examples are meant to dispel some common confusions.

Examples 8.4 *a) Consider the Laplace experiment described by the sample space* $\Omega \overset{\text{def}}{=} \{1,2,\ldots,8\}$ *and the probabilities* $P(i) \overset{\text{def}}{=} \frac{1}{8}$ *for all* $i \in \Omega$. *Define the events*

$$
A_1 \overset{\text{def}}{=} \{1,2,3,4\} \, , \qquad A_2 \overset{\text{def}}{=} \{1,2,5,6\} \, , \qquad A_3 \overset{\text{def}}{=} \{3,4,5,6\} \, .
$$

Then, one easily checks that the events A_1, A_2 *and* A_3 *are pairwise independent, but*

$$
P(A_1 \cap A_2 \cap A_3) = 0 \neq \frac{1}{8} = P(A_1)P(A_2)P(A_3) \, .
$$

Thus, while independence of events $A_1, \ldots, A_m$ clearly implies their pairwise in-dependence, the opposite is not true.

b) Consider now rolling a single dice. The sample space is $\Omega \overset{def}{=} \{1, 2, \ldots, 6\}$ and each elementary event $\{i\}$ has probability $P(i) \overset{def}{=} \frac{1}{6}$. Define the events

$$A_1 \overset{def}{=} \{1, 2, 3\}, \qquad A_2 \overset{def}{=} \{2, 4, 6\}, \qquad A_3 \overset{def}{=} \{1, 2, 4, 5\}.$$

Again it is easy to check that

$$P(A_1 \cap A_2 \cap A_3) = \frac{1}{6} = P(A_1)P(A_2)P(A_3).$$

Furthermore, $P(A_1 \cap A_3) = P(A_1)P(A_3)$ and $P(A_2 \cap A_3) = P(A_2)P(A_3)$ hold but

$$P(A_1 \cap A_2) = \frac{1}{6} \neq \frac{1}{4} = P(A_1)P(A_3).$$

Thus, for $A_1, \ldots, A_m$ to be independent it does not suffice to require

$$P(\cap_{j=1}^m A_j) = \prod_{j=1}^m P(A_j).$$

8.1.3 Independent Collections of Events

We say that the collections of events $C_1, C_2, \ldots, C_m$ are *independent* if for each choice $A_i \in C_i$, the events $A_1, \ldots, A_m$ are independent.

The following result establishes the equivalence between independence of partitions and independence of the corresponding algebras.

Proposition 8.5 *Let $P_1, \ldots, P_m$ be partitions of Ω. The following statements are equivalent:*

a) *the partitions $P_1, \ldots, P_m$ are independent;*

b) *the algebras $A(P_1), \ldots, A(P_m)$ are independent;*

c) *for every choice of $A_i \in P_i$ we have*

$$P(\cap_{j=1}^m A_j) = \prod_{j=1}^m P(A_j).$$

d) *For every choice of $A_i \in A(P_i)$ we have*

$$P(\cap_{j=1}^m A_j) = \prod_{j=1}^m P(A_j).$$

Proof We start by proving that d) implies b). Take any set $J \subset \{1, \ldots, m\}$ and take for each $j \in J$ an arbitrary $A_j \in \mathcal{A}(\mathcal{P}_j)$. Setting $A_i \stackrel{\text{def}}{=} \Omega$ for each $i \notin J$ we get

$$\cap_{j \in J} A_j = \cap_{j=1}^m A_j \; .$$

Thus,

$$P(\cap_{j \in J} A_j) = P(\cap_{j=1}^m A_j) = \prod_{j=1}^m P(A_j) = \prod_{j \in J} P(A_j) \; ,$$

where in the second equality we used assumption d) and in the last equality that $P(A_j) = P(\Omega) = 1$ for $j \notin J$. This proves that b) holds.

That b) implies a) is clear since $\mathcal{P}_i \subset \mathcal{A}(\mathcal{P}_i)$ holds. Moreover, that a) implies c) is immediate from the definition of the independence of $\mathcal{P}_1, \ldots, \mathcal{P}_m$.

It remains to prove that c) implies d). Assume that $\mathcal{P}_i$ is given by

$$\mathcal{P}_i = \{A_1^i, \ldots, A_{r_i}^i\} \; .$$

Take now any $A_j \in \mathcal{A}(\mathcal{P}_j)$. Since A_j is the union of atoms of $\mathcal{P}_j$ we find sets $I_j \subset \{1, \ldots, r_j\}$ such that

$$A_j = \cup_{k_j \in I_j} A_{k_j}^j \; .$$

Note that this union is a disjoint union. We then have

$$
\begin{aligned}
P(A_1 \cap \ldots \cap A_m) &= P([\cup_{k_1 \in I_1} A_{k_1}^1] \cap \ldots \cap [\cup_{k_m \in I_m} A_{k_m}^m]) \\
&= P(\cup_{k_1 \in I_1} \cdots \cup_{k_m \in I_m} [A_{k_1}^1 \cap \ldots \cap A_{k_m}^m]) \\
&= \sum_{k_1 \in I_1} \cdots \sum_{k_m \in I_m} P(A_{k_1}^1 \cap \ldots \cap A_{k_m}^m) \\
&= \sum_{k_1 \in I_1} \cdots \sum_{k_m \in I_m} P(A_{k_1}^1) \cdot \ldots \cdot P(A_{k_m}^m) \\
&= \sum_{k_1 \in I_1} P(A_{k_1}^1) \cdot \ldots \cdot \sum_{k_m \in I_1} P(A_{k_m}^m) \\
&= P(A_1) \cdot \ldots \cdot P(A_m) \; .
\end{aligned}
$$

This implies independence of $\mathcal{A}(\mathcal{P}_1), \ldots, \mathcal{A}(\mathcal{P}_m)$ and concludes the proof of the proposition. □

Example 8.6 *Assume we roll a die twice. The corresponding sample space is*

$$\Omega \stackrel{\text{def}}{=} \{(1,1), \ldots, (1,6), \ldots, (6,1), \ldots, (6,6)\}$$

and every elementary event (i, j) has the same probability $P(i, j) = \frac{1}{36}$. Let A_j be the event "rolling the first time j" and B_i the event "rolling the second time i", i.e.

$$A_j = \{(j, 1), \ldots, (j, 6)\} \qquad and \qquad B_i = \{(1, i), \ldots, (6, i)\} \; .$$

Since $A_j \cap B_i = \{(j,i)\}$ holds, we have $P(A_j) = P(B_i) = \frac{6}{36}$ and $P(A_j \cap B_i) = \frac{1}{36}$. Therefore, $P(A_j \cap B_i) = P(A_j)P(B_i)$ and, thus, A_j and B_i are independent. Therefore, the partitions

$$\mathcal{P} \stackrel{def}{=} \{A_1, \ldots, A_6\} \qquad and \qquad \mathcal{Q} \stackrel{def}{=} \{B_1, \ldots, B_6\}$$

are independent.

8.2   Independence of Random Variables

If X is a random variable, we have previously introduced the natural partition $\mathcal{P}(X)$ of X.

The random variables $X_1, \ldots, X_m$ are said to be independent if their natural partitions $\mathcal{P}(X_1), \ldots, \mathcal{P}(X_m)$ — or, equivalently (Proposition 8.5) , their natural algebras $\mathcal{A}(X_1), \ldots, \mathcal{A}(Y_m)$ — are independent.

We next prove some equivalent formulations of independence among random variables which should help clarify the concept.

Proposition 8.7 *The following statements are equivalent:*

a) *$X_1, \ldots, X_m$ are independent.*

b) *For each $(v_1, \ldots, v_m) \in \mathbb{R}^m$ the sets $[X_1 = v_1], \ldots, [X_m = v_m]$ satisfy*

$$P([X_1 = v_1, \ldots, X_m = v_m]) = P([X_1 = v_1]) \ldots P([X_m = v_m]) .$$

c) *For any family $A_1, \ldots, A_m$ of subsets of $\mathbb{R}$ the sets $[X_1 \in A_1], \ldots, [X_m \in A_m]$ satisfy*

$$P([X_1 \in A_1, \ldots, X_m \in A_m]) = P([X_1 \in A_1]) \ldots P([X_m \in A_m]) . \qquad (8.1)$$

Proof By definition, $X_1, \ldots, X_m$ are independent if and only if $\mathcal{A}(X_1), \ldots, \mathcal{A}(X_m)$ are independent. Note that

$$[X_1 \in A_1, \ldots, X_m \in A_m] = [X_1 \in A_1] \cap \ldots \cap [X_m \in A_m] ,$$

and that by Remark 7.11, the set $[X_i \in A_i]$ belongs to $\mathcal{A}(X_i)$ for $1 \leq i \leq m$. It follows from Proposition 8.5 that (8.1) holds. It follows that condition a) implies condition c).

Setting $A_i \stackrel{def}{=} \{v_i\}$, it is clear that condition d) implies condition b).

Finally, each atom in $\mathcal{P}(X_i)$ can be written as $[X_i = v_i]$ for a suitable number v_i. Hence, b) implies a).   □

Proposition 8.8 *The random variables $X_1, \ldots, X_m$ are independent if and only if for each choice $J_1, \ldots, J_r$ of disjoint subsets of $\{1, \ldots, m\}$, we have that the partitions*

$$\mathcal{P}(X_{j_1}; j_1 \in J_1), \ldots, \mathcal{P}(X_{j_r}; j_1 \in J_r)$$

are independent.

Proof Take events $B_i \in \mathcal{P}(X_{j_i}; j_i \in J_i)$. Then,

$$B_i = \cap_{j_i \in J_i} A_{j_i}$$

holds for suitable $A_{j_i} \in \mathcal{P})(X_{j_i})$. Since, $\mathcal{P}(X_1), \ldots, \mathcal{P}(X_m)$ are independent, we can use Proposition 8.3 to get that

$$P(B_1 \cap \ldots \cap B_r) = P(B_1) \cdot \ldots \cdot P(B_r)$$

holds. From Proposition 8.5 we obtain that $\mathcal{P}(X_{j_1}; j_1 \in J_1), \ldots, \mathcal{P}(X_{j_r}; j_1 \in J_r)$ are independent partitions. $\qquad\square$

Example 8.9 *Assume we roll a die twice. Recall the corresponding sample space*

$$\Omega \overset{def}{=} \{(1,1), \ldots, (1,6), \ldots, (6,1), \ldots, (6,6)\}$$

and that every elementary event (i,j) has the same probability $P(i,j) = \frac{1}{36}$. Define the random variables X and Y by setting for each $(i,j) \in \Omega$

$$X(i,j) \overset{def}{=} i \qquad and \qquad Y(i,j) \overset{def}{=} j$$

Then, the standard partitions are

$$\mathcal{P}(X) = \{A_1, \cdots, A_6\} \qquad with \ A_k \overset{def}{=} \{(k,1), \ldots, (k,6)\}$$

and

$$\mathcal{P}(Y) = \{B_1, \cdots, B_6\} \qquad with \ B_k \overset{def}{=} \{(1,k), \ldots, (6,k)\} \,,$$

which as we have already seen in a previous example are independent. Thus, X and Y are independent.

Measurability and independence

It is to be expected that if two random variables X and Y are measurable with respect to independent partitions $\mathcal{P}$ and $\mathcal{Q}$, respectively, then X and Y will also be independent.

Proposition 8.10 *Assume that $\mathcal{P}_1, \ldots, \mathcal{P}_m$ are independent partitions and that $X_1, \ldots, X_m$ are random variables. If X_i is $\mathcal{P}_i$-measurable for each $i \in \{1, \ldots, m\}$, then $X_1, \ldots, X_m$ are independent random variables.*

Proof This is clear since $\mathcal{P}_i$-measurability of X_i implies that $\mathcal{A}(X_i) \subset \mathcal{A}(\mathcal{P}_i)$. Since $\mathcal{A}(\mathcal{P}_1), \ldots \mathcal{A}(\mathcal{P}_m)$ are independent, the same follows for $\mathcal{A}(X_1), \ldots, \mathcal{A}(X_m)$, i.e. $X_1, \ldots, X_m$ are independent random variables. $\qquad\square$

Functions of Independent Random Variables

Proposition 8.11 *Let $X_1, \ldots, X_m$ be independent random variables and assume that $m_1, \ldots, m_r$ are natural numbers such that $m_1 + \ldots + m_r = m$. Set $m_0 \stackrel{def}{=} 1$ and let $f_i : \mathbb{R}^{m_i+1-m_i} \to \mathbb{R}$ be given functions. Then*

$$
\begin{aligned}
Y_1 &\stackrel{def}{=} f_1(X_1, \ldots, X_{m_1}), \\
Y_2 &\stackrel{def}{=} f_2(X_{m_1+1}, \ldots, X_{m_1+m_2}),
\end{aligned}
$$

$$\vdots$$

$$\vdots$$

$$
Y_r \stackrel{def}{=} f_r(X_{m_1+\ldots+m_{r-1}+1}, \ldots, X_{m_r})
$$

are also independent.

Proof From Proposition 7.20 we know that Y_i is $\mathcal{Q}_i$-measurable, where we have set

$$
\mathcal{Q}_i \stackrel{def}{=} P(X_{m_1+\ldots+m_{i-1}+1}, \ldots, X_{m_1+\ldots+m_i}).
$$

By Proposition 8.8, we have that $\mathcal{Q}_1, \ldots, \mathcal{Q}_r$ are independent partitions. Thus, by Proposition 8.10 the random variabes $Y_1, \ldots, Y_r$ are independent. $\qquad\square$

8.3 Expectations, Variance and Independence

An important result is the following which tells us that independent random variables are also uncorrelated.

Proposition 8.12 *Let $X_1, \ldots, X_m$ be independent random variables. Then,*

$$
E_P[X_1 \cdot \ldots \cdot X_m] = E_P[X_1] \cdot \ldots \cdot E_P[X_m].
$$

Proof Let X and Y be two independent random variables. Assume that

$$
X(\Omega) = \{x_1, \ldots, x_r\} \quad \text{and} \quad Y(\Omega) = \{y_1, \ldots, y_s\}.
$$

We can then write

$$
X = x_1 1_{A_1} + \ldots + x_r 1_{A_r} \quad \text{and} \quad Y = y_1 1_{B_1} + \ldots + y_s 1_{A_s}
$$

where we have set $A_i \stackrel{def}{=} [X = x_i]$ and $B_j = [Y = y_j]$. Moreover,

$$
\mathcal{P}(X) = \{A_1, \ldots, A_r\} \quad \text{and} \quad \mathcal{P}(Y) = \{B_1, \ldots, A_s\}.
$$

Since X and Y are independent we have

$$
E_P[1_{A_i} 1_{B_j}] = E_P[1_{A_i \cap B_j}] = P(A_i \cap B_j) = P(A_i)P(B_j) = E_P[1_{A_i}]E_P[1_{B_j}].
$$

We thus get

$$E_P[XY] \;=\; \sum_i \sum_j E_P[1_{A_i} 1_{B_j}] = \sum_i \sum_j E_P[1_{A_i}] E_P[1_{B_j}]$$
$$=\; E_P[X] E_P[Y] \,.$$

It follows that the proposition holds for two independent random variables. Assuming that it holds for m random variables we will proceed by induction to show that it must hold for $m + 1$.

By Proposition 8.11 we know that X_{m+1} and $Y \stackrel{\text{def}}{=} X_1 \cdot \ldots \cdot X_m$ are independent random variables. By what we have just shown we get

$$E_P[X_1 \cdot \ldots \cdot X_m \cdot X_{m+1}] = E_P[Y X_{m+1}] = E_P[Y] E_P[X_{m+1}] \,.$$

But $E_P[X_1 \cdot \ldots \cdot X_m] = E_P[X_1] \cdot \ldots \cdot E_P[X_m]$ holds by induction hypothesis, so

$$E_P[X_1 \cdot \ldots \cdot X_m \cdot X_{m+1}] = E_P[X_1] \cdot \ldots \cdot E_P[X_m] E_P[X_{m+1}]$$

holds, proving the validity of the product formula for $m + 1$ independent random variables. □

If two random variables are independent, we have

$$Cov_P(X, Y) = E_P[XY] - E_P[Y] E_P[Y] = E_P[Y] E_P[Y] - E_P[Y] E_P[Y] = 0 \,.$$

Hence, they are uncorrelated. Thus, using Proposition 5.18 we obtain the following result.

Corollary 8.13 *Independent random variables X and Y are uncorrelated, i.e.*

$$Cov_P(X, Y) = 0 \,.$$

In particular, if $X_1, \ldots, X_r$ is a collection of uncorrelated random variables, we have

$$Var_P\left(\sum_{i=1}^{r} X_i\right) = \sum_{i=1}^{r} Var_P(X_i) \,.$$

8.4   Sequences of Independent Experiments

A frequent situation one encounters is that a random experiment consists in the consecutive execution of several independent experiments. More formally assume we are given N random experiments which are independent of each other. Assume furthermore that we have probabilistic models for each of them which we denote by

$$(\Omega_i, P_i) = \{\omega_1^i, \ldots, \omega_{n_i}^i\}$$

for $1 \leq i \leq N$. The natural sample space Ω for the model of our combined experiment is of course the cartesian product of $\Omega_1, ..., \Omega_N$

$$\Omega \overset{\text{def}}{=} \Omega_1 \times \cdots \times \Omega_N .$$

A typical element of Ω is thus given by

$$\omega = (\omega_{j_1}^1, \cdots, \omega^N j_N) ,$$

with $\omega_{j_i}^i \in \Omega_i$ representing the outcome of the i-th experiment.

We now intuitively derive two conditions that a probability measure P on Ω should satisfy if these experiments are to be modelled as being independent of each other. Let

$$\omega = (\omega_{j_1}^1, \ldots, \omega_{j_N}^N) \in \Omega$$

be an arbitrary elementary event in Ω. Set $A_{j_i} \overset{\text{def}}{=} \{\omega_{j_i}^i\}$ and denote by $\hat{A}_{j_i}$ the event in Ω that $\{\omega_{j_i}^i\}$ occurred in the i-th experiment. Formally, this event is given by

$$\hat{A}_{j_i} \overset{\text{def}}{=} \Omega_1 \times \cdots \times \Omega_{j-1} \times A_{j_i} \times \Omega_{j+1} \times \cdots \times \Omega_N .$$

The independence assumption means that the outcome in the i-th experiment should be independent of the outcome of the other experiments. In particular we would expect that the probability of $\hat{A}_{j_i}$ occurring should be the same as the probability of A_{j_i} occurring, i.e. we should have

$$P(\hat{A}_{j_i}) = P(\Omega_1 \times \cdots \times \Omega_{j-1} \times A_{j_i} \times \Omega_{j+1} \times \cdots \times \Omega_N) = P_j(A_{j_i}) = P_i(\omega_{j_i}^i). \quad (8.2)$$

Furthermore, we should have that the events $\hat{A}_{j_1}, \ldots, \hat{A}_{j_N}$ are independent, i.e. noting that $\{\omega\} = \{(\omega_{j_1}^1, \ldots, \omega_{j_N}^N)\} = \hat{A}_{j_1} \cap \ldots \cap \hat{A}_{j_N}$ holds, we should require

$$\begin{aligned}
P(\omega_{j_1}^1, \ldots, \omega_{j_N}^N) &= P(\omega) \\
&= P(\hat{A}_{j_1}) \times \cdots \times P(\hat{A}_{j_N}) \\
&= P(A_{j_N}) \times \cdots \times P(A_{j_N}) \\
&= P_1(\omega_{j_1}^1) \times \cdots \times P_N(\omega_{j_N}^N) .
\end{aligned}$$

Therefore, if P is a probability measure on Ω such that events $\hat{A}^1, \ldots, \hat{A}^N$ are independent it must be unique, since a probability measure is fully determined by the values it takes on the elementary events.

In fact we can use the above expression to define P. We obviously get:

Lemma 8.14 *Set for each* $\omega = (\omega_{j_1}^1, \ldots, \omega_{j_N}^N) \in \Omega$,

$$P(\omega) \overset{\text{def}}{=} P(\omega_{j_1}^1) \cdot \ldots \cdot P(\omega_{j_N}^N),$$

and define for each $A \subset \Omega$,

$$P(A) \overset{\text{def}}{=} \sum_{\omega \in A} P(\omega).$$

Then, P defines a probability measure on Ω. It is the only probability measure on Ω satisfying

$$P(A_1 \times \ldots \times A_N) = P(A_1) \cdot \ldots \cdot P(A_N),$$

for all $A_1 \times \ldots \times A_N \subset \Omega$.

8.4.1   Bernoulli Experiments

Let a Bernoulli experiment be given by the sample space

$$\Omega_0 \overset{\text{def}}{=} \{0, 1\}$$

and the probabilities

$$P_0(1) = p \in (0, 1) \qquad \text{and} \qquad P_0(0) = 1 - p.$$

We interpret the outcome 1 as being a success and 0 as a failure. Consider now the situation where we repeat the experiment N times and the outcome of each repetition is independent of the outcomes of the other repetitions. The sample space will consist of the set

$$\Omega \overset{\text{def}}{=} \underbrace{\Omega_0 \times \cdots \times \Omega_0}_{N \text{ times}}.$$

The probability of an elementary event $\omega = (\omega_1, \omega_2, \ldots, \omega_N) \in \Omega$ will be given by

$$P(\omega) = P_0(\omega_1) \cdot \ldots \cdot P_0(\omega_N).$$

Define now for each $i = 1, 2, \ldots, N$ the random variable $Z_i : \Omega \to \mathbb{R}$ by setting

$$Z_i(\omega) \overset{\text{def}}{=} \omega_i.$$

Thus, Z_i keeps track of whether the outcome of the i-th repetition was a success $(Z_i(\omega) = 1)$ or a failure $(Z_i(\omega) = 0)$.

Lemma 8.15 *The random variables $Z_1, \ldots, Z_N$ are independent. Moreover,*

$$P([Z_i = \delta]) = p^\delta (1 - p)^{1 - \delta}$$

and, therefore,

$$E_P[Z_i] = p \qquad \text{and} \qquad Var_P[Z_i] = p(1 - p).$$

Proof For $1 \leq i \leq N$ choose $\delta_i \in \{0, 1\}$. Note that

$$[Z_i = \delta] = \Omega \times \cdots \times \Omega \times \{\delta\} \times \Omega \times \cdots \times \Omega \, ,$$

where $\{\delta\}$ appears in the i-th position. But these events are independent by construction of P. It follows that $Z_1, \ldots, Z_N$ are independent.
Moreover, we have

$$P([Z_i = \delta]) = P_0(\Omega) \cdot \ldots \cdot P_0(\Omega) \cdot P_0(\delta) \cdot P_0(\Omega) \cdot \ldots \cdot P_0(\Omega) = P_0(\delta) = p^\delta (1-p)^{1-\delta} \, ,$$

where $P(\delta)$ appears in the i-th position.
Expected value and variance of Z_i can now be calculated directly.   □

Define now for each $0 \leq t \leq N$ the random variables $D_t : \Omega \to \mathbb{R}$ defined by

$$D_t(\omega) \stackrel{\text{def}}{=} Z_1(\omega) + \ldots + Z_t(\omega) \, .$$

Thus, D_t counts the number of successes in the outcome ω up to time t .
By means of the random variable D_N we can give a nice representation for the probability measure on Ω.

Lemma 8.16

$$P(\omega) = p^{D_N(\omega)} (1-p)^{N - D_N(\omega)}$$

Proof Note that

$$\{\omega\} = [Z_1 = \omega_1] \cap \ldots \cap [Z_N = \omega_N] \, .$$

By the preceding lemma, $Z_1, \ldots, Z_N$ are independent and

$$
\begin{aligned}
P(\omega) &= P([Z_1 = \omega_1] \cap \ldots \cap [Z_N = \omega_N]) \\
&= P([Z_1 = \omega_1]) \cdot \ldots \cdot P([Z_N = \omega_N]) \\
&= p^{Z_1(\omega)}(1-p)^{1 - Z_1(\omega)} \cdot \ldots \cdot p^{Z_N(\omega)}(1-p)^{1 - Z_N(\omega)} \\
&= p^{D_N(\omega)}(1-p)^{N - D_N(\omega)} \, .
\end{aligned}
$$

□

From Lemma 8.15 and using Corollary 8.13 we immediately get expected value and variance for each of the D_t.

Lemma 8.17 *For any $0 \leq t \leq N$ we have:*

$$E_P[D_t] = tp \, ,$$

and

$$Var_P[D_t] = tp(1-p) \, .$$

9.1.1   The Underlying Probability Structure

The starting point for our probabilistic model consists of three components:

- A sample space

$$\Omega = \{\omega_1, \omega_2, \ldots, \omega_n\}$$

  representing all the possible final states of the world at time T.

- An information structure

$$\mathcal{I} = \{\mathcal{P}_0, \mathcal{P}_1, \ldots, \mathcal{P}_T\}$$

  describing the arrival of information with the passage of time. The partitions are assumed to be given by

$$\mathcal{P}_t = \{A_1^t, \ldots, A_{r_t}^t\}$$

  with $r_0 = 1 \leq r_2 \leq \ldots \leq r_{T-1} \leq r_T = n$.

- A probability measure

$$P : \Omega \to [0, 1]$$

  describing the "natural" probabilities for the possible states of the world, i.e. $p_i \stackrel{\text{def}}{=} P(\omega_i)$ represents the probability that at time $t = T$ the world will be revealed to be in state ω_i.

9.1.2   The Space of Alternatives and Contingent Claims

In contrast to the single-period case in a multi-period market we can consider financial contracts maturing not only at time T but also at any of the intermediate dates $1 \leq S \leq T$.

A *(European) alternative* with maturity S is a pattern of (positive or negative) payments at time S which is contingent on the information revealed at that time. It can be therefore represented by a non-negative $\mathcal{P}_S$-measurable random variable $X_S : \Omega \to \mathbb{R}$. Since, X_S is constant on the atoms $A_1^S, \ldots, A_{r_S}^S$ of $\mathcal{P}_S$, it follows that X_S can be represented by the vector

$$(X(A_1^S), \ldots, X(A_{r_S}^S)) \in \mathbb{R}^{r_S} ,$$

where r_S obviously denotes the number of atoms of $\mathcal{P}_S$. We interpret $X_S(A_i^S)$ as the amount to be paid or received if at time S it transpires that the event A_i^S occurred. The space of European alternatives with maturity S is denoted by $\mathcal{A}_S$, i.e.

$$\mathcal{A}_S \stackrel{\text{def}}{=} L(\Omega, \mathcal{P}_S) .$$

Any non-negative alternative with maturity S will be called a *(European) contingent claim* [1]. The cone of contingent claims maturing at S is denoted by

$$\mathcal{A}_S^+ \overset{\text{def}}{=} \{X_S \in \mathcal{A}_S \ ; \ X_S \geq 0\} \ .$$

9.1.3 The Securities

In our multi-period economy, $(N + 1)$ (non-dividend-paying) non-zero contingent claims with maturity T will be traded. In contrast to the single-period case, these securities can be bought and sold for a price at any of the trading dates $t = 0, \ldots, T$. As in the single-period case these securities will be called the *basic securities*, or just *securities* for short.

What does it take to describe a security?

In order to describe a security we will need to specify:

- its payoff $S_T : \Omega \to \mathbb{R}^+$ at time $t = T$; and

- its price $S_t : \Omega \to \mathbb{R}$ at each time t, which since it must be observable at time t has to be a $\mathcal{P}_t$-measurable random variable.

Thus, we need to specify a $(\mathcal{P}_t)$-adapted process (S_t) which we call its *price process*. Since securities entitle the holder to a non-zero payoff at maturity we assume that

$$S_t(\omega) > 0$$

holds for all ω such that $S_T(\omega) > 0$, i.e. their prices should be positive as long as the prospect of getting a non-zero positive payoff at maturity exists. In particular, we have

$$S_0 > 0 \ .$$

The $(N + 1)$ price processes

For $i = 0, 1, \ldots, N$ the price process of the i-th security,

$$(S_t^i)_{0 \leq t \leq T} \ ,$$

is $(\mathcal{P}_t)$-adapted and non-negative and satisfies

$$S_t^i(\omega) > 0$$

for all ω such that $S_T^i(\omega) > 0$. In particular,

$$S_0^i > 0$$

holds for all i.

---

[1]The adjective "European" is used to distinguish these claims from so-called "American" claims to be studied in Chapter 16. Since in this chapter we will not encounter any other alternatives than European, we shall drop the qualifier "European" and just speak of alternatives.

9.2.1 Static and Dynamic Portfolios

A *(static) portfolio* is just a combination of positions in the different securities available in our economy. It can therefore be represented by an $N + 1$-tuple

$$\Phi = (\phi^0, \phi^1, \ldots, \phi^N)$$

where ϕ^j represents the position in the j-th security. Note that a negative entry for ϕ^i entails having sold short $|\phi^i|$ units of the i-th security.

Intermediate trading allows the possibility of *re-balancing* the portfolio at intermediate dates, i.e. of liquidating certain positions and building up others. This is captured by the concept of a dynamic portfolio.

A *dynamic portfolio* maturing at time S is a portfolio whose composition changes over time up to time S. It can thus be represented by a sequence $\Phi = (\Phi_t)_{1 \leq t \leq S}$ of portfolios. One thinks about a dynamic portfolio in the following way:

- At time $t = 0$ the agent buys a portfolio Φ_1 at the prices prevailing at time $t = 0$.

- At time $t = 1$ he liquidates the portfolio Φ_1 and sets up a new portfolio Φ_2. Both transactions are done at the prices prevailing at time $t = 1$.

- Generally speaking, the agent holds the portfolio Φ_t during the period $[t-1, t]$ and liquidates it at time t at the then prevailing prices, simultaneously setting up the new portfolio Φ_{t+1}. This is done at time $S - 1$ for the last time. The portfolio Φ_S is then held until time $t = S$, the date at which it is liquidated and the proceeds consumed.

9.2.2 Trading Strategies

In reality the re-balancing of portfolios at intermediate dates will occur in the light of the information available at that time. In order to model this we introduce the notion of a trading strategy.

Designing a trading strategy entails prescribing at time $t = 0$ how to re-balance the portfolio at each date t contingent on the information available at that time. Mathematically, a *trading strategy* maturing at time S is a stochastic dynamic portfolio

$$\Phi = (\Phi_t)_{1 \leq t \leq S} = \left(\phi_t^0, \ldots, \phi_t^N \right)_{1 \leq t \leq S}$$

which is *$\mathcal{I}$-predictable*, i.e.

a) for each $i = 1, \ldots, N$, the stochastic process $(\phi_t)_{1 \leq t \leq S}$ is predictable with respect to $(\mathcal{P}_t)_{1 \leq t \leq S}$, i.e. ϕ_t^i is $\mathcal{P}_{t-1}$-measurable for each $i = 0, 1, \ldots, S$ and $t = 1, 2, \ldots, S$.

The requirement of $\mathcal{I}$-predictability ensures that Φ_t, the composition of the portfolio to be held during the period $[t - 1, t]$, will be known at time $t - 1$, as soon as the information $\mathcal{P}_{t-1}$ arrives. This is natural since Φ_t has to be set up at time $t - 1$.

9.2.3 Acquisition and Liquidation Value of a Trading Strategy

If $\mathbf{\Phi} = (\Phi_t)_{1 \leq t \leq S}$ is a trading strategy maturing at time S its *acquisition value* at time $t = 0, \ldots, S - 1$ is defined by

$$V_t^{Acq}[\mathbf{\Phi}] \stackrel{\text{def}}{=} \phi_{t+1}^0 S_t^0 + \phi_{t+1}^1 S_t^1 + \ldots + \phi_{t+1}^N S_t^N .$$

The *liquidation value* of $\mathbf{\Phi}$ at time t is defined as

$$V_t^{Liq}[\mathbf{\Phi}] \stackrel{\text{def}}{=} \phi_t^0 S_t^0 + \phi_t^1 S_t^1 + \ldots + \phi_t^N S_t^N .$$

We call $V_0^{Acq}[\mathbf{\Phi}]$ the *initial value* and $V_S^{Liq}[\mathbf{\Phi}]$ the *terminal value* of $\mathbf{\Phi}$. Note that the liquidation value at time t corresponds to the value at which the portfolio Φ_t is liquidated at time t just before acquiring the new portfolio Φ_{t+1} at the value $V_t^{Acq}[\mathbf{\Phi}]$.

9.2.4 Self-Financing Strategies

There is a special class of trading strategies which we would like to single out: self-financing trading strategies. In general when re-balancing within a trading strategy, the difference between the proceeds obtained when liquidating one portfolio and the funds needed for setting up the portfolio for the next period of time may be positive (allowing the withdrawal of funds for consumption) or negative (requiring an injection of funds). *Self-financing trading strategies* are strategies $\mathbf{\Phi} = (\Phi_t)_{0 \leq t \leq S}$ for which the proceeds obtained when liquidating Φ_t at time t are equal to the amount of money needed for setting up Φ_{t+1}, i.e. strategies for which

$$V_t^{Liq}[\mathbf{\Phi}] = V_t^{Acq}[\mathbf{\Phi}] .$$

Having set up portfolio Φ_1, a self-financing trading strategy finances each re-balancing on its own: injections or withdrawals are neither necessary nor possible. For a self-financing strategy we need not distinguish between its acquisition and the liquidation value. We therefore define its *value* to be

$$V_t[\mathbf{\Phi}] \stackrel{\text{def}}{=} V_t^{Liq}[\mathbf{\Phi}] = V_t^{Acq}[\mathbf{\Phi}] .$$

The set of self-financing strategies is closed under the basic operations as is captured in the next remark.

Remark 9.1 *The sum of two self-financing strategies with maturity S is again a self-financing trading strategy with maturity S. Multiplying a self-financing strategy with maturity S by any scalar generates a new self-financing trading strategy with maturity S.*

9.3   Attainability and Replication

The question of replication and attainability is slightly more subtle for multi-period economies. An alternative X_S maturing at time S is said to be *attainable* if there exists a self-financing trading strategy maturing $\mathbf{\Phi} = (\Phi_t)_{0 \leq t \leq S}$ at time S, such that

$$V_S[\mathbf{\Phi}] = X_S$$

holds, i.e. such that the terminal value of the trading strategy coincides with the contingent claim. Any strategy with this property is called a *replicating, hedging* or *generating* strategy for X_S.

We set

$$\mathcal{M}_S \overset{\text{def}}{=} \{X_S \in \mathcal{A}_S \, ; X_S \text{ is attainable}\} \, .$$

9.4   The Law of One Price and Linear Pricing Functionals

In this section, as we did for the the single-period model, we investigate conditions under which there is a well-defined linear pricing functional (the Law of One Price). The next section will be devoted to conditions under which this functional is strongly positive (absence of arbitrage).

9.4.1   Prices of Attainable Alternatives

Assume X_S is an attainable alternative with maturity S, i.e. $X_S \in \mathcal{M}_S$, i.e. we can find a replicating portfolio $\mathbf{\Phi}$. This means that $\mathbf{\Phi}$ is a self-financing strategy such that $V_S[\mathbf{\Phi}] = X_S$. It would make sense to define the fair value X_S by the cost of setting up a replicating strategy, i.e. by

$$\pi_{0,S}(X_S) \overset{\text{def}}{=} V_0[\mathbf{\Phi}] \, .$$

This is of course only meaningful if every replicating strategy has the same initial value, i.e. if the *Law of One Price* holds.

Lemma 9.2 *The Law of One Price holds for the maturity S if and only if each strategy $\mathbf{\Phi}$ replicating $Y_S \equiv 0$ satisfies $V_0[\mathbf{\Phi}] = 0$.*

Proof Since the zero strategy, i.e. $\Phi_t \equiv 0$, obviously replicates the zero-claim for any maturity, the Law of One Price implies that any strategy replicating the zero-claim with maturity S must have zero initial value.

Assume that the Law of One Price does not hold. Then we find two self-financing strategies $\mathbf{\Phi}$ and $\mathbf{\Psi}$ with maturity S such that $V_S[\mathbf{\Phi}] = V_S[\mathbf{\Psi}]$ but $V_0[\mathbf{\Phi}] \neq V_0[\mathbf{\Psi}]$. Then, the self-financing strategy $\mathbf{\Phi} - \mathbf{\Psi}$ replicates the zero-claim maturing at S and $V_0[\mathbf{\Phi} - \mathbf{\Psi}] \neq 0$.   □

As in the single-period case, we show that either the Law of One Price holds for alternatives maturing at time S, or for each such alternative we find replicating portfolios with an arbitrarily pre-specified initial value.

Proposition 9.3 *One of the following alternatives holds:*

   a) The Law of One Price *holds for alternatives maturing at time S.*

   b) For any alternative $X_S \in \mathcal{M}_S$ and each $\lambda \in \mathbb{R}$ we find a replicating strategy $\boldsymbol{\Phi}$ for X_S satisfying

$$V_0[\boldsymbol{\Phi}] = \lambda .$$

Proof Assume the Law of One Price does not hold. We then find a replicating strategy $\boldsymbol{\Phi}$ for the zero-claim maturing at time S with non-zero initial value. Let X_S be an attainable and $\boldsymbol{\Psi}$ a corresponding replicating strategy. Then, for each $\delta \in \mathbb{R}$ the strategy $\boldsymbol{\Psi} + \delta\boldsymbol{\Phi}$ replicates X_S. Choose $\delta \stackrel{\text{def}}{=} \frac{\lambda - V_0[\boldsymbol{\Phi}]}{V_0[\boldsymbol{\Phi}]}$. Then,

$$V_S[\boldsymbol{\Psi} + \delta\boldsymbol{\Phi}] = V_S[\boldsymbol{\Psi}] + \delta V_S[\boldsymbol{\Phi}] = V_S[\boldsymbol{\Psi}] = X_S ,$$

and $V_0[\boldsymbol{\Psi} + \delta\boldsymbol{\Phi}] = \lambda$. □

The next result states that the Law of One Price holds for all maturities if and only if it holds for the longest maturity T.

Proposition 9.4 *The Law of one Price holds for all maturities if and only if it holds for alternatives maturing at time T.*

Proof Let $\boldsymbol{\Phi} = (\boldsymbol{\Phi}_t)_{0 \leq t \leq S}$ be a replicating strategy for the zero-claim maturing at S. Define now

$$\tilde{\boldsymbol{\Phi}}_t \stackrel{\text{def}}{=} \begin{cases} \boldsymbol{\Phi}_t & \text{if } t \leq S, \\ 0 & \text{if } S < t \leq T . \end{cases}$$

Then, $\tilde{\boldsymbol{\Phi}} = (\tilde{\boldsymbol{\Phi}}_t)_{0 \leq t \leq T}$ is a self-financing strategy replicating the zero-claim with maturity T. Since the Law of One Price holds for this maturity we have

$$V_0[\boldsymbol{\Phi}] = V_0[\tilde{\boldsymbol{\Phi}}] = 0 .$$

It follows that the Law of One Price holds for the maturity S. □

When we say "the Law of One Price holds", we will always mean the Law of One Price holds for all maturities.

Assume the Law of One Price holds. In this case we can define for each attainable alternative X_S maturing at time S its *fair value* by setting

$$\pi_{0,S}(X_S) \stackrel{\text{def}}{=} V_0[\boldsymbol{\Phi}]$$

for some replicating strategy $\boldsymbol{\Phi}$. Hence, as in the single-period case, the fair value is the amount of money needed to start up any of the self-financing strategies replicating X_S. The following result tells us that the mapping which assigns to each attainable alternative its fair value is linear.

Theorem 9.5 *Assume the Law of One Price holds. Then, $\pi_{0,S} : \mathcal{M}_S \to \mathbb{R}$ is a linear functional for all $S = 1, \ldots, T$.*

Proof Take any two attainable alternatives X_S and Y_S in $\mathcal{M}_S$ and corresponding replicating strategies $\mathbf{\Phi}_X$ and $\mathbf{\Phi}_Y$. If $\lambda \in \mathbb{R}$ is given, then $\mathbf{\Phi}_X + \lambda\mathbf{\Phi}_Y$ is a replicating portfolio for $X + \lambda Y$. Therefore:

$$
\begin{aligned}
\pi_{0,S}(X + \lambda Y) &= V_0(\mathbf{\Phi}_X + \lambda\mathbf{\Phi}_Y) = V_0(\mathbf{\Phi}_X) + \lambda V_0(\mathbf{\Phi}_Y) \\
&= \pi_{0,S}(X) + \lambda\pi_{0,S}(Y).
\end{aligned}
$$

$\square$

In markets where the Law of One Price holds, we will call $\pi_{0,S} : \mathcal{M}_S \to \mathbb{R}$, the *pricing functional* for maturity S. By a slight abuse of notation we will drop the subscript S and just write π_0 instead of $\pi_{0,S}$. This should lead to no confusion.

9.5   No-Arbitrage and Strongly Positive Pricing Functionals

We now turn to the condition which ensures that the pricing functional is strongly positive.

Admissible strategies and arbitrage

An *admissible trading strategy* maturing at time S is a self-financing strategy $\mathbf{\Phi} = (\mathbf{\Phi}_t)_{0 \leq t \leq S}$, whose value is non-negative at all times, i.e.

$$
V_t(\mathbf{\Phi}) \geq 0 ,
$$

holds for all $t \in \{0, 1, \ldots, S\}$. We are now able to formalize the notion of an arbitrage opportunity in the context of a multi-period model.

An *arbitrage opportunity* or *arbitrage trading strategy* with maturity S is an admissible strategy $\mathbf{\Phi} = (\mathbf{\Phi}_t)_{0 \leq t \leq S}$ with zero initial value $V_0(\mathbf{\Phi})$ and non-zero terminal value $V_S(\mathbf{\Phi})$.

Observe, that since an arbitrage opportunity has zero initial value there is no need to invest any money at all at inception. On the other hand, since the strategy is self-financing, neither is there a need to invest any money at any time: re-balancing will always be financed by liquidating the previous portfolio. Since the strategy is admissible, at no time will the holder have any liabilities which he would not be able to honor if called upon. In particular, the terminal value will be non-negative so that in no state of the world will liquidation of $\mathbf{\Phi}_S$ result in a loss. Finally, since the terminal value is non-zero, there will be a state of the world where liquidating $\mathbf{\Phi}_S$ will result in a positive gain. We thus see that the above definition captures the intuitive notion of an arbitrage opportunity as a possibility of having at zero-cost a potential gain and no potential losses.

Remark 9.6 *We just observe that, as in the one-period model, underlying price processes admitting no arbitrage correspond to prices in economic equilibrium. Arbitrage opportunities create an incentive to trade. If arbitrage opportunities exist, at given prices, demand will exceed supply leading to disequilibrium.*

Proposition 9.7 *The market admits no arbitrage opportunities for all maturities if and only if it admits no arbitrage opportunities for the maturity T.*

Proof Let Φ be an arbitrage opportunity for the maturity S. Define,

$$\tilde{\Phi}_t \stackrel{\text{def}}{=} \begin{cases} \Phi_t & \text{if } t \leq S, \\ (V_S[\Phi], 0, \ldots, 0) & \text{if } S < t \leq T \,. \end{cases}$$

Then it is clear that $\tilde{\Phi}$ is an arbitrage opportunity for the maturity T. Hence, if the market does not admit arbitrage opportunities for the maturity T, it cannot admit arbitrage opportunities for lesser maturities either.

The converse, is of course, trivial.   □

When we say "the market admits no arbitrage opportunities", we will always mean the market admits no arbitrage opportunities for any maturity.

No-arbitrage and value processes

Recall that we describe the flow of information by means of the information structure $\mathcal{I} = \{\mathcal{P}_0, \ldots, \mathcal{P}_T\}$, where the partition $\mathcal{P}_t$ is given by

$$\mathcal{P}_t = \{A_1^t, \ldots, A_{r_t}^t\} \,.$$

Let X_S be an attainable alternative maturing at time S and $\Phi = (\Phi_t)_{0 \leq t \leq S}$ a corresponding replicating strategy. Recall that $V_t[\Phi]$ is $\mathcal{P}_t$-measurable. Therefore, when at time t we learn to which of the atoms $A_1^t, \ldots, A_{r_t}^t$ the true outcome belongs, we will be able to evaluate $V_t[\Phi]$.

Assume now that A_j^t has been revealed to have occurred and that X_S is non-negative on A_j^t, i.e. knowing that A_j^t has occurred we also know that the payoff of the claim will be non-negative for sure. In this situation we expect that the value of Φ at time t is non-negative on A_j^t. This is proved in the next lemma.

Lemma 9.8 *Let $\Phi = (\Phi_t)_{0 \leq t \leq S}$ be a self-financing strategy maturing at S and assume that for some $0 \leq s_0 < S$ and $1 \leq j \leq r_{s_0}$ we have*

$$V_S[\Phi] \geq 0 \qquad \text{on } A_j^{s_0} \,.$$

Then,

$$V_t[\Phi] \geq 0 \qquad \text{on } A_j^{s_0} \,,$$

for all $s_0 \leq t \leq S$.

Moreover, if

$$V_S[\mathbf{\Phi}] = 0 \qquad on \ A_j^{so}$$

holds, then,

$$V_t[\mathbf{\Phi}] = 0 \qquad on \ A_j^{so} \ ,$$

for all $s_0 \le t \le S$.

Proof Assume the first statement does not hold and set

$$t_0 \overset{\text{def}}{=} \sup\{t; 0 \le t < S \text{ and } P([V_t[\mathbf{\Phi}] < 0] \cap A_j^{so}) > 0\} \ .$$

Then,

- $s_0 \le t_0 < S$;
- $P([V_{t_0}[\mathbf{\Phi}] < 0] \cap A_j^{so}) > 0$; and
- $V_t[\mathbf{\Phi}] \ge 0$ on A_j^{so} for all $t_0 < t \le S$.

Set now

$$A \overset{\text{def}}{=} [V_{t_0}[\mathbf{\Phi}] < 0] \cap A_j^{so} \ .$$

Observe that $V_{t_0}[\mathbf{\Phi}]$ is $\mathcal{P}_{t_0}$-measurable and, thus, also $\mathcal{P}_{t_0}$-measurable. It follows that $[V_{t_0}[\mathbf{\Phi}] < 0]$ belongs to $\mathcal{A}(\mathcal{P}_{t_0})$. Thus, it follows that A also belongs to $\mathcal{A}(\mathcal{P}_{t_0})$. Define now a self-financing strategy $\tilde{\mathbf{\Phi}} = (\tilde{\mathbf{\Phi}}_t)_{0 \le t \le S}$ by setting

$$\tilde{\mathbf{\Phi}}_t \overset{\text{def}}{=} \begin{cases} 0 & \text{if } t \le t_0, \\ 1_A[\Phi_t - (\frac{V_{t_0}[\mathbf{\Phi}]}{S_{t_0}^0}, 0, \dots, 0)] & \text{if } t_0 < t \le S \ . \end{cases}$$

Since $A \in \mathcal{A}(\mathcal{P}_{t_0})$ we have that $\tilde{\mathbf{\Phi}}$ is predictable. Moreover,

$$V_t[\tilde{\mathbf{\Phi}}] = \begin{cases} 0 & \text{if } t \le t_0, \\ 1_A[V_t[\mathbf{\Phi}] - \frac{V_{t_0}[\mathbf{\Phi}]}{S_{t_0}^0} S_t^0] & \text{if } t_0 < t \le S \ . \end{cases}$$

In particular, for $t_0 < t \le S$ we have

$$V_t[\tilde{\mathbf{\Phi}}] = 0 \text{ on } A^c \quad \text{and} \quad V_t[\tilde{\mathbf{\Phi}}] > 0 \text{ on } A; \ .$$

Thus, $\tilde{\mathbf{\Phi}}$ is an arbitrage opportunity for maturity S, contradicting the absence of arbitrage opportunities in this market.

Assume now that $V_S[\mathbf{\Phi}] = 0$ on A_j^{so}. By what we have just proved we get $V_t[\mathbf{\Phi}] \ge 0$ for $s_0 \le t \le S$. On the other hand, $-\mathbf{\Phi}$ also satisfies $V_S[-\mathbf{\Phi}] = 0$, so that $V_t[\mathbf{\Phi}] \le 0$ must hold. It follows that $V_t[\mathbf{\Phi}] = 0$, as asserted. $\qquad\qquad\qquad \square$

It is worthwhile stating a particular instance of this lemma separately. It tells us that any strategy replicating a contingent claim (a non-negative alternative) must be admissible .

Corollary 9.9 *Assume the absence of arbitrage opportunities. Then if* $\mathbf{\Phi}$ *is a self-financing strategy with* $V_S(\mathbf{\Phi}) \geq 0$, *then* $V_t(\mathbf{\Phi}) \geq 0$ *for all* $0 \leq t \leq S$. *If* $V_S(\mathbf{\Phi}) = 0$ *holds, then* $V_t(\mathbf{\Phi}) = 0$ *for all* $0 \leq t \leq S$.

Proof Choose $s_0 = 0$ and $A_j^{s_0} = \Omega$ in Lemma 9.8. $\qquad\square$

As a corollary we get that if no arbitrage opportunities exist, the Law of One Price holds in an even stronger sense: at each time the values of two replicating strategies for the same alternative have the same value process.

Theorem 9.10 *Assume the absence of arbitrage opportunities. Then if two self-financing strategies* $\mathbf{\Phi}$ *and* $\mathbf{\Psi}$ *satisfy* $V_S(\mathbf{\Phi}) = V_S(\mathbf{\Psi})$ *for some* S, *then*

$$V_t(\mathbf{\Phi}) = V_t(\mathbf{\Psi})$$

for $0 \leq t \leq S$.

Proof Let $\mathbf{\Phi}$ and $\mathbf{\Psi}$ be two strategies replicating the attainable alternative X_S maturing at S. Then, $\mathbf{\Phi} - \mathbf{\Psi}$ replicates the zero-claim maturing at S. It follows from Corollary 9.9 that $V_t[\mathbf{\Phi} - \mathbf{\Psi}] = 0$, and thus $V_t[\mathbf{\Phi}] = [\mathbf{\Psi}]$for all $0 \leq t \leq S$. $\quad\square$

Strong positivity of the pricing functional

Since the Law of One Price holds, we can consider the linear pricing functional

$$\pi_0 : \mathcal{M}_S \to \mathbb{R} \,,$$

given by

$$\pi_0(X_S) \stackrel{\text{def}}{=} V_0[\mathbf{\Phi}]$$

for any strategy $\mathbf{\Phi}$ replicating X_S.
The following result relates absence of arbitrage to strong positivity of the pricing functional.

Theorem 9.11 *The market admits no arbitrage opportunities if and only if the pricing functional is strongly positive.*

Proof Assume the market does not admit arbitrage opportunities. Let X_S be a non-zero positive claim and $\mathbf{\Phi} = (\mathbf{\Phi}_t)_{0 \leq t \leq S}$ a corresponding replicating strategy. By Corollary 9.9 we have that $\mathbf{\Phi}$ is admissible, i.e. $V_t[\mathbf{\Phi}] \geq 0$ for $0 \geq t \geq S$. If $V_0[\mathbf{\Phi}] = 0$ holds, $\mathbf{\Phi}$ would be an arbitrage opportunity, since $V_S[\mathbf{\Phi}] = X_S > 0$. Thus,

$$\pi_0[X_S] = V_0[\mathbf{\Phi}] > 0 \,,$$

and $\pi_0 : \mathcal{M}_S \to \mathbb{R}$ is strongly positive.

Proposition 10.2 *Let a random variable X be given. For any partition $\mathcal{P} = \{B_1, B_2, \ldots, B_m\}$, its conditional expectation $E_P[X|\mathcal{P}]$ is $\mathcal{P}$-measurable. Moreover, it is the only $\mathcal{P}$-measurable random variable Y satisfying*

$$E_P[Y1_{B_i}] = E_P[X1_{B_i}] \tag{10.2}$$

for all i.

Proof Set $Z \stackrel{\text{def}}{=} E_P[X|\mathcal{P}]$. As a linear combination of the $\mathcal{P}$-measurable random variables $1_{B_1}, \ldots, 1_{B_m}$ it is clear that Z is also $\mathcal{P}$-measurable. We now show that Z satisfies (10.2). For any i we have

$$Z1_{B_i} = E_P[X|B_i]1_{B_i} \ .$$

Thus, using Proposition 10.1 we obtain

$$E_P[Z1_{B_i}] = E_P[X|B_i]E_P[1_{B_i}] = \frac{1}{P(B_i)}E_P[X1_{B_i}]P(B_i) = E_P[X1_{B_i}] \ ,$$

as claimed.

Let Y be any other $\mathcal{P}$-measurable random variable satisfying (10.2). The $\mathcal{P}$ measurability of Z and Y imply that they are constant on atoms of $\mathcal{P}$. From this and using (10.2) we infer that

$$Z(B_i) = \frac{E_P[Z1_{B_i}]}{P(B_i)} = \frac{E_P[Y1_{B_i}]}{P(B_i)} = Y(B_i)$$

holds for all i. It follows that $Z = Y$ and, therefore, that Z is the only $\mathcal{P}$-measurable random variable satisfying (10.2). $\qquad\square$

Remarks 10.3   *a) By Proposition 10.1 we see that (10.2) is equivalent to*

$$E_P[Y|B_i] = E_P[X|B_i]$$

for all i.

b) Since $\mathcal{A}(\mathcal{P})$ consists of unions of atoms of $\mathcal{P}$ we immediately see that (10.2) is equivalent to

$$E_P[Y1_A] = E_P[X1_A]$$

for $A \in \mathcal{A}(\mathcal{P})$.

An important consequence of the above result is the following.

Corollary 10.4 *Let a random variable X and a partition $\mathcal{P} = \{B_1, B_2, \ldots, B_m\}$ be given. Then*

$$E_P[E_P[X|\mathcal{P}]] = E_P[X] \ .$$

Proof By definition

$$
\begin{aligned}
E_P[E_P[X|\mathcal{P}]] &= E_P[E_P[X|B_1]1_{B_1} + \ldots + E_P[X|B_m]1_{B_m}] \\
&= E_P[X|B_1]E_P[1_{B_1}] + \ldots + E_P[X|B_m]E_P[1_{B_m}] \\
&= P(B_1)E_P[X|B_1] + \ldots + P(B_m)E_P[X|B_m] \\
&= E_P[X] \,,
\end{aligned}
$$

where for the last step we used Proposition 10.1. □

10.1.3 Properties of the Conditional Expectation

Let a partition $\mathcal{P} = \{B_1, B_2, \ldots, B_m\}$ of Ω be given. In this section we list some important properties of conditional expectations with respect to $\mathcal{P}$. We start with its linearity, whose easy proof is left to the reader.

Proposition 10.5 *The operation of taking conditional expectations is linear, i.e.*

$$
E_P[X + \lambda Y|\mathcal{P}] = E_P[X|\mathcal{P}] + \lambda E_P[Y|\mathcal{P}]
$$

holds for random variables X and Y and any scalar λ.

The properties we list next will be repeatedly used in the remainder of the book.

Proposition 10.6 *a) If X is $\mathcal{P}$-measurable, then*

$$
E_P[X|\mathcal{P}] = X \,,
$$

i.e. on arrival of $\mathcal{P}$ we know the exact value of X and do not need to take any expectations.

b) If Z is any $\mathcal{P}$-measurable random variable, we have:

$$
E_P[ZX|\mathcal{P}] = ZE_P[X|\mathcal{P}] \,,
$$

i.e. we may "take out" what is "known".

c) Taking conditional expectations preserves positivity, i.e. if X is a positive random variable, then $E_P[X|\mathcal{P}]$ is also positive.

d) If $\mathcal{Q}$ is a partition finer than $\mathcal{P}$, i.e. if every atom of $\mathcal{P}$ is contained in an atom of $\mathcal{Q}$, then

$$
E_P[E_P[X|\mathcal{P}]|\mathcal{Q}] = E_P[X|\mathcal{Q}] \,.
$$

This is sometimes called the tower property.

Similarly, $(X_t)_{0 \leq t \leq T}$ is said to be a *P-supermartingale* with respect to $\mathbb{P}$ if

$$E_P[X_t | \mathcal{P}_s] \leq X_s$$

holds for all $0 \leq s \leq t \leq T$. Assume the price process of a security is a supermartingale. After the arrival of the information $\mathcal{P}_s$ we will then expect a price decrease with respect to the prevailing price $X_s(\omega)$. Supermartingales are thus associated with "unfavorable" games, i.e. games where wealth is expected to decrease.

Remark 10.13 *Note that the properties of being a martingale, submartingale or supermartingale depend crucially not only on the probability measure P but also on the information structure $\mathcal{P}$.*

10.3.2   Maxima and Minima

Taking maxima of submartingales, respectively minima of supermartingales, generates a new submartingale, respectively a new supermartingale.

Lemma 10.14 *Let $(X_t^1), \ldots, (X_t^m)$ be a sequence of submartingales. Then, the process (Z_t^{max}) defined by*

$$Z_t^{max} \stackrel{def}{=} \max\{X_t^1, \ldots, X_t^m\}$$

is a submartingale.
If $(X_t^1), \ldots, (X_t^m)$ is a sequence of supermartingales, the process (Z_t^{min}) defined by

$$Z_t^{min} \stackrel{def}{=} \min\{X_t^1, \ldots, X_t^m\}$$

is a supermartingale.

Proof From Corollary 7.13 we know that (Z_t^{max}) is $(\mathcal{P}_t)$-adapted. Assume $s \leq t$, then

$$E_P[X_t^i] \geq X_s^i$$

for all $i = 1, \ldots, m$. This implies

$$\max\{E_P[X_t^1], \ldots, E_P[X_t^m]\} \geq \max\{X_s^1, \ldots, X_s^m\} = Z_s^{max} .$$

Using, Lemma 10.7 we obtain

$$E_P[Z_t^{max}] = E_P[\max\{X_t^1, \ldots, X_t^m\}] \geq \max\{E_P[X_t^1], \ldots, E_P[X_t^m]\} \geq Z_s^{max} .$$

The assertion for supermartingales is proved similarly.   □

10.3.3   Doob's Decomposition

Any adapted stochastic process can be decomposed into the sum of a martingale and a predictable process. A simple corollary is that any sub- or supermartingale can be decomposed into the sum of a martingale and a predictable decreasing, respectively increasing, process. In the following we assume that a probability measure P together with an information structure $\mathcal{P}$ is fixed.

Theorem 10.15 *Any adapted process $(X_t)_{0 \leq t \leq T}$ has a unique decomposition*

$$X_t = M_t + A_t \, ,$$

where (M_t) is a martingale and (A_t) is a predictable process with $A_0 = 0$.
If (X_t) is a sub- or a supermartingale, then the predictable process (A_t) is increasing, respectively decreasing.

Proof For $t = 0, \ldots, T$ define M_t and A_t recursively by

$$M_0 := X_0 \, , \ A_0 := 0$$

and

$$M_{t+1} := M_t + X_{t+1} - E[X_{t+1}|\mathcal{F}_t] \, , \ A_{t+1} := A_t - X_t + E[X_{t+1}|\mathcal{F}_t] \, .$$

By conditioning with respect to $\mathcal{F}_t$ we obtain

$$E[M_{t+1}|\mathcal{F}_t] = M_t$$

i.e. M_t is indeed a martingale. It is clear from the definition that A_{t+1} is $\mathcal{F}_t$-measurable, i.e. (A_t) is a predictable process. To verify the uniqueness of the decomposition let

$$X_t = M'_t + A'_t$$

be a second decomposition with the same properties. We proceed inductively. Clearly

$$X_0 = M_0 = M'_0 \quad \text{and} \quad A_0 = A'_0 = 0 \, .$$

The induction hypothesis is $M_t = M'_t$ and $A_t = A'_t$. By conditioning the identity

$$X_{t+1} = M_{t+1} + A_{t+1} = M'_{t+1} + A'_{t+1}$$

with respect to $\mathcal{F}_t$, and by using the martingale property, we find that

$$M_t + E[A_{t+1}|\mathcal{F}_t] = M'_t + E[A'_{t+1}|\mathcal{F}_t] \, .$$

Chapter 11

The Fundament
of Asset Pricing

The most important implication of the c
tence of a positive linear pricing rule, u
finite state spaces is the same as the ex
that correctly price all assets. Taken tog
refer collectively to these results as the F
Pricing.

This chapter is devoted to the proof of the Fir
rems of Asset Pricing, which characterize the e
lent martingale measures in terms of absence of
respectively. We mimic the approach adopted i
Compared to the one-period case, the only add
the proof of the correspondence between strict
ing functional and equivalent martingale measu
slightly more involved in some technical aspect

We assume that an economy as described in Ch
notation of that chapter. We begin with the obv
of an equivalent martingale measure to the mul

Using the induction hypothesis $M_t = M$
we obtain

$$E[A_{t+1}|\mathcal{F}_t] = A_{t+1} =$$

and thus $M_{t+1} = M'_{t+1}$, which completes
If (X_t) is a supermartingale $X_t - E[X_{t+1}$
Analogously (A_t) is increasing if (X_t) is a

In the Doob decomposition the process M
shock and A_t is called the *predictable pa*
(supermartingale) can be decomposed int
shock .

Concluding Remarks and
Reading

Since the seminal work of J.M. Harrison an
and S.R, Pliska ([27]) martingales have p
cal finance. Equivalent martingale measure
chapter have proved to be a powerful tool i
Some useful references are [36], [49], [37], a

11.1   Change of Numeraire and Discounting

In general, we are interested in relative prices, i.e. in having a key for expressing a number of units of a given security in terms of a number of units of another security. Up to now, we have chosen to do this by implicitly assuming the existence of a currency in which we have expressed all prices. However, there is no reason why not to express prices relative to one of the available securities, i.e. choosing this security as a *numeraire* or *accounting unit*. For a security to qualify as a numeraire we need only require that it has a strongly positive price process, i.e. that at each time and in each state it has a strictly positive price.

Our assumptions on the 0-th security ensure that we can choose it as a numeraire. The price of the j-th security *discounted by* or *relative to* the 0-th security is defined by

$$\tilde{S}_t^j \stackrel{\text{def}}{=} \frac{S_t^j}{S_t^0} \; .$$

For any trading strategy $\mathbf{\Phi} = (\Phi_t)_{0 \leq t \leq S}$ we have the corresponding notion of value relative to the new numeraire. Its *discounted liquidation value* at time t is given by

$$\tilde{V}_t^{Liq}[\mathbf{\Phi}] \stackrel{\text{def}}{=} \phi_t^0 \tilde{S}_t^0 + \phi_t^1 \tilde{S}_t^1 + \ldots + \phi_t^N \tilde{S}_t^N = \frac{V_t^{Liq}[\mathbf{\Phi}]}{S_t^0} \; .$$

The *discounted acquisition value* of $\mathbf{\Phi}$ at time t is defined by

$$\tilde{V}_t^{Acq}[\mathbf{\Phi}] \stackrel{\text{def}}{=} \phi_{t+1}^0 \tilde{S}_t^0 + \phi_{t+1}^1 \tilde{S}_t^1 + \ldots + \phi_{t+1}^N \tilde{S}_t^N = \frac{V_t^{Acq}[\mathbf{\Phi}]}{S_t^0} \; .$$

Obviously, for a self-financing strategy $\mathbf{\Phi}$, discounted acquisition and liquidation values coincide and we can write

$$\tilde{V}_t[\mathbf{\Phi}] \stackrel{\text{def}}{=} \tilde{V}_t^{Liq}[\mathbf{\Phi}] = \tilde{V}_t^{Acq}[\mathbf{\Phi}] = \frac{V_t[\mathbf{\Phi}]}{S_t^0} \; .$$

Finally, for an alternative X_S maturing at time S we define the *discounted alternative*, $\tilde{X}_S$, by

$$\tilde{X}_S \stackrel{\text{def}}{=} \frac{X_S}{S_S^0} \; .$$

If X_S is attainable and $\mathbf{\Phi}$ a corresponding replicating strategy, we will obviously have

$$\tilde{V}_S[\mathbf{\Phi}] = \tilde{X}_S \; .$$

> For the remainder of this chapter we will adopt the 0-th security as a numeraire. Discounted prices, values and alternatives will always be understood with respect to this security.

11.2 Martingales and Asset Prices

A probability measure Q on Ω is said to be an *equivalent martingale measure* for the economy if the following two conditions are satisfied:

- Q is *equivalent* to P, i.e. for all $i = 1, \ldots, n$

$$Q(\omega_i) > 0 .$$

- The discounted price process $(\tilde{S}_t^j)_{0 \le t \le T}$ is a Q-martingale with respect to $\mathcal{I}$ for all $j = 0, 1, \ldots, N$, i.e.

$$E_Q[\tilde{S}_t^j | \mathcal{P}_s] = \tilde{S}_s^j$$

for all $s \le t \le T$ and $j \in \{1, \ldots, N\}$.

In particular, taking $s = 0$ in the Q-martingale condition, we get

$$S_0^j = S_0^0 \, E_Q[\tilde{S}_t^j | \mathcal{P}_0] = S_0^0 \, E_Q[\tilde{S}_t^j].$$

Thus, having an equivalent martingale measure Q at our disposal we are able to recover the price S_0^j of each basic security by taking the expected values of discounted future prices, where the expected values are taken with respect to Q.

11.2.1 Equivalent Martingale Measures and Self-Financing Trading Strategies

Assume there are no arbitrage opportunities and let X_S be an attainable alternative maturing at time S. Denote by $\mathbf{\Phi} = (\Phi_t)_{0 \le t \le S}$ a replicating strategy, i.e. an admissible trading strategy such that

$$V_S[\mathbf{\Phi}] = X_S .$$

Recall that the fair value of the alternative X_S was defined as the cost of starting up the self-financing strategy $\mathbf{\Phi}$, i.e.

$$\pi_0(X) = V_0[\mathbf{\Phi}] .$$

Proposition 11.1 *Let Q be a measure equivalent to P. Then, Q is an equivalent martingale measure if and only if for all self-financing strategies $\mathbf{\Phi} = (\Phi_t)_{0 \le t \le S}$ with maturities $0 \le S \le T$, the discounted value process, $(\tilde{V}_t[\mathbf{\Phi}])_{0 \le t \le S}$, is a Q-martingale with respect to $\mathcal{I}$, i.e.*

$$E_Q[\tilde{V}_t[\mathbf{\Phi}] | \mathcal{P}_s] = \tilde{V}_s[\mathbf{\Phi}] \qquad (11.1)$$

holds for all $s \le t \le S$.

Since $q_i > 0$, $i = 1, \ldots, n$ it follows that Q defines a probability measure on Ω which is equivalent to P. Then (11.7) is equivalent to

$$E_Q\left[V_T[\boldsymbol{\Phi}]\right] = \sum_{i=1}^{n} q_i V_T[\boldsymbol{\Phi}](\omega_i) = \sum_{i=1}^{n} \frac{\alpha_i}{\sum_{j=1}^{n} \alpha_j} V_T[\boldsymbol{\Phi}](\omega_i) = 0 , \qquad (11.8)$$

for any predictable and self-financing strategy ϕ satisfying $V_0[\boldsymbol{\Phi}] = 0$.

From Proposition 11.3 and Remark 11.4 we obtain that Q is an equivalent martingale measure. $\hfill\square$

11.3   The Fundamental Theorems of Asset Pricing

Exactly as in the one-period setting of Chapter 6, the correspondence between strongly positive extensions of the pricing functional and equivalent martingale measures together with our result on strongly positive extensions of the pricing functional (Proposition 9.14) allows us to immediately deduce the two fundamental theorems of asset pricing.

The First Fundamental Theorem of Asset Pricing reads as follows.

Theorem 11.7 *The existence of an equivalent martingale measure is equivalent to the absence of arbitrage opportunities.*

This is a very satisfying theorem since it tells us that what might be considered as desirable from a purely computational point of view — the existence of a martingale measure — is in fact guaranteed by the absence of arbitrage, a purely economic requirement.

The above result says that a martingale measure exists if there are no arbitrage opportunities. It does not tell us, however, how many of them exist. The uniqueness question is settled by the Second Fundamental Theorem of Asset Pricing which we state next.

Theorem 11.8 *Assume that the market admits no arbitrage opportunities. Then, there exists a unique equivalent martingale measure if and only if the market is complete.*

11.4   Risk-Adjusted and Forward-Neutral Measures

It is important to note that reference to the choice of a numeraire — i.e. the security with respect to which we discount all other assets — is essential when talking about equivalent martingale measures. Different numeraires will give rise to different equivalent martingale measures.

In Chapter 6 we introduced two possible variants for the 0-th security: A money-market account or a zero-bond. In absence of arbitrage and under completeness, these two variants give rise to two different equivalent martingale measures. They are called the *risk-neutral probability measure* and the *forward-neutral probability measure*, respectively.

Risk-neutral probability measure   Taking the money-market account as the numeraire we obtain the risk-adjusted measure. Noting that for the money-market account we have $S_0^0 = 1$ we obtain

$$\pi(X) = V_0(\mathbf{\Phi}) = E_Q[\frac{V_T(\mathbf{\Phi})}{S_T^0}|\mathcal{P}_0] = E_Q[\frac{X}{S_T^0}] .$$

Forward-neutral probability measure   If we take the zero-bond as the numeraire we obtain the forward-neutral measure. For the zero-bond we have $S_T^0 = 1$. Hence,

$$\pi(X) = S_0^0 \tilde{V}_0(\mathbf{\Phi}) = S_0^0 E_Q[\tilde{V}_T(\mathbf{\Phi})|\mathcal{P}_0] = S_0^0 E_Q[V_T(\mathbf{\Phi})|\mathcal{P}_0] = S_0^0 E_Q[X].$$

We thus see that the forward-neutral measure is more tractable for discounting payments at maturity T.

In models with deterministic interest rates we had remarked that the money-market account and the zero-bond were essentially the same. Thus, in this case, the risk-adjusted and the forward-neutral probabilities are also equal.

Concluding Remarks and Suggestions for Further Reading

The Fundamental Theorems of Asset Pricing were proved in [26] and [27] for discrete time and continuous time. The term Fundamental Theorems of Asset Pricing was coined in [22]. The impact of these theorems has been incredible even in the world of practitioners. For general versions of these theorems see [19].

Other treatments of this material can be found in the first chapter of the very elegant book [39] or in [20] and [47].

Chapter 12

The Cox–Ross–Rubinstein Model

This rounds up the principal conclusion of this paper: the simple two-state process is really the essential ingredient of option pricing by arbitrage methods. This is surprising, perhaps, given the mathematical complexity of some of the current models in this field. But it is reassuring to find such simple economic arguments at the heart of this powerful theory

J.C. Cox, S.A. Ross and M. Rubinstein

In this chapter we show how to apply the fundamental theorems of asset pricing to a simple but important example. The Cox–Ross–Rubinstein model is a multi-period generalization of the one-period model considered in Chapter 2. Building on this model and the central limit theorem, Chapter 14 will provide a complete derivation of the celebrated Black–Scholes option pricing formula.

12.1   The Cox–Ross–Rubinstein Economy

We consider an N-period economy. During each of the periods the development of the economy can be either "good" or "bad". We will assume that the development of the economy during any given period is independent of the past developments, i.e. we shall assume that there exists a number $p \in (0, 1)$ such that for any period the probability of the development being "good" is equal to p. Consequently the probability of a "bad" development is given by $1 - p$.

Proof

$$
\begin{aligned}
S(t) &= S_0 \left(1 + y_g\right)^{D_t} \left(1 + y_b\right)^{t-D_t} \\
&= S_0 \left(1 + y_g\right)^{D_{t-1}+Z_t} \left(1 + y_b\right)^{[t-1-D_{t-1}]+[1-Z_t]} \\
&= S_0 \left(1 + y_g\right)^{D_{t-1}} \left(1 + y_b\right)^{t-1-D_{t-1}} \left(1 + y_g\right)^{Z_t} \left(1 + y_b\right)^{1-Z_t} \\
&= S_{t-1} \left(1 + y_g\right)^{Z_t} \left(1 + y_b\right)^{1-Z_t} .
\end{aligned}
$$

$\square$

The return of the stock for the period t is the random variable

$$
R_t : \Omega \to \{1 + y_g, 1 + y_b\}
$$

given by

$$
R_t(\omega) \stackrel{\text{def}}{=} \frac{S_t(\omega)}{S_{t-1}(\omega)} = (1 + y_g)^{Z_t(\omega)} (1 + y_b)^{1-Z_t(\omega)} , \tag{12.4}
$$

as immediately seen from (12.3). The following lemma captures an important feature of the Cox–Ross–Rubinstein model.

Lemma 12.2 *The random variables R_t , $t = 1, \dots, N$ are independent.*

Proof The assertion follows from the independence of the random variables Z_t and Proposition 8.11 by observing that

$$
R_t = g(Z_t)
$$

with $g(s) \stackrel{\text{def}}{=} (1 + y_g)^s (1 + y_b)^{1-s}$. $\square$

Stock prices and information

The following simple result allows us to identify the information up to time t with the observations of prices of the risky security up to that time. It will repeatedly prove useful.

Lemma 12.3 *If X is a $\mathcal{P}_t$-measurable random variable, there exists a function $F : \mathbb{R}^t \to \mathbb{R}$ such that*

$$
X(\omega) = F(S_0, S_1(\omega), \dots, S_t(\omega))
$$

holds for all $\omega \in \Omega$.

Proof Since X is $\mathcal{P}_t$-measurable it is constant on the atoms $[Z_1 = a_1, \dots, Z_t = a_t]$ of the partition $\mathcal{P}_t$. Define

$$
f(a_1, \dots, a_t) \stackrel{\text{def}}{=} X([Z_1 = a_1, \dots, Z_t = a_t]) .
$$

Then obviously

$$X(\omega) = f(Z_1(\omega), \ldots, Z_t(\omega)) \ .$$

Since

$$S_t = S_{t-1} (1 + y_g)^{Z_t} (1 + y_b)^{1 - Z_t}$$

holds, it follows that

$$Z_t = g(S_{t-1}, S_t) \ ;$$

where

$$g(x, y) \stackrel{\text{def}}{=} \log[\frac{y}{x(1 + y_b)^N}] \bigg/ \log[\frac{1 + y_g}{1 + y_b}] \ .$$

Hence, Z_t is a function of S_t and S_{t-1} and therefore X a function of the random variables $S_0, S_1, \ldots, S_t$, proving the assertion of the lemma. $\quad\square$

12.2 Parametrizing the Model

In typical applications of the Cox–Ross–Rubinstein model we are given a time interval $[0, T]$ during which one wishes to model the behavior of a particular stock or stock index. For a given $N \in \mathbb{N}$ one considers an N-period Cox–Ross–Rubinstein economy where trading is admitted at times

$$0, \frac{T}{N}, 2\frac{T}{N}, \ldots, (N-1)\frac{T}{N}, T \ .$$

We can re-scale time by choosing the unit $\frac{T}{N}$. Hence, time t with respect to re-scaled time will correspond to time $t\frac{T}{N}$ in the original time scaling. Instead of the notations Z_t and D_t from the previous section we use the obvious $Z_{N,t}$ and $D_{N,t}$. The economy is fully specified by the risk-free rate r_N which determines the evolution of the money market account

$$B_{N,t} = (1 + r_N)^t$$

and the parameters $y_{b,N}$ and $y_{g,N}$ which determine the stock price process

$$S_{N,t} = S_0 (1 + y_{g,N})^{D_{N,t}} (1 + y_{b,N})^{t - D_{N,t}}.$$

The natural probability P_N given by $0 < p_N < 1$ also needs to be specified. In this section we show how these parameters can be chosen to fit historical observations.

The risk-free rate

Let $R > 0$ be such that e^{-TR} is the price of a zero-bond with face value 1 and maturing at time T. In our N-period model the value of such a bond would be $\frac{1}{(1+r_N)^N}$. Hence, for consistency, it is natural to set

$$(1 + r_N)^N = e^{TR}$$

or, equivalently,

$$1 + r_N = e^{\frac{T}{N}R} .$$

The number R is usually called the *continuous compounding rate*.

The stock

Assume first that $y_{b,N}, y_{g,N}$ and p_N are given and write

$$S_{N,T}(\omega) = S_0 e^{H_N(\omega)}$$

with

$$H_N(\omega) \stackrel{\text{def}}{=} \log \frac{S_{N,T}(\omega)}{S_0(\omega)} = D_{N,N}(\omega) \log(1 + y_{g,N}) + (N - D_{N,N}(\omega)) \log(1 + y_{b,N}).$$

Note that because $D_{N,N}$ is Bernoulli distributed with parameters N and p_N we have

$$
\begin{aligned}
E_{P_N}[H_N] &= N p_N \log(1 + y_{g,N}) + N(1 - p_N) \log(1 + y_{b,N}) , & (12.5)\\
\text{Var}_{P_N}[H_N] &= N p_N (1 - p_N)[\log(1 + y_{g,N}) - \log(1 + y_{b,N})]^2 . & (12.6)
\end{aligned}
$$

On the other side, from historical data it is possible to estimate

$$\mu \stackrel{\text{def}}{=} \frac{1}{T} E_{P_N}[H_N] = \frac{1}{T} E_{P_N}[\log \frac{S_{N,N}}{S_0}] ,$$

$$\sigma^2 \stackrel{\text{def}}{=} \frac{1}{T} Var_{P_N}[H_N] = \frac{1}{T} Var_{P_N}[\log \frac{S_{N,N}}{S_0}].$$

Thus, it is natural to choose $y_{g,N}, y_{b,N}$, and p_N such that $T\mu$ and $T\sigma^2$ match (12.5) and (12.6), respectively. By requiring this we obtain the following two equations:

$$p_N \log(1 + y_{g,N}) + N(1 - p_N) \log(1 + y_{b,N}) = \frac{T}{N}\mu , \tag{12.7}$$

$$N p_N (1 - p_N)[\log(1 + y_{g,N}) - \log(1 + y_{b,N})]^2 = \frac{T}{N}\sigma^2 . \tag{12.8}$$

We thus have two equations but three unknowns. We introduce the assumption that a "good" development followed by a "bad" development leaves the price of the stock unchanged, and vice versa, i.e.

$$(1 + y_{g,N})(1 + y_{b,N}) = 1$$

or, equivalently,

$$\log(1 + y_{g,N}) = -\log(1 + y_{b,N}) \ . \tag{12.9}$$

Using this equality, (12.7) and (12.8) translate into

$$(2p_N - 1)\log(1 + y_{g,N}) \quad = \quad \frac{T}{N}\mu \ , \tag{12.10}$$

$$4p_N(1 - p_N)\log^2(1 + y_{g,N}) \quad = \quad \frac{T}{N}\sigma^2 \ . \tag{12.11}$$

Squaring (12.10) and adding the result to (12.11) we easily get

$$\log(1 + y_{g,N}) = \sqrt{\frac{T}{N}\sigma^2 + \frac{T^2}{N^2}\mu^2}$$

and therefore, by (12.9),

$$\log(1 + y_{b,N}) = -\sqrt{\frac{T}{N}\sigma^2 + \frac{T^2}{N^2}\mu^2}.$$

Equivalently stated, we get

$$1 + y_{g,N} = e^{\sqrt{\frac{T}{N}\sigma^2 + \frac{T^2}{N^2}\mu^2}} \ , \tag{12.12}$$

$$1 + y_{b,N} = e^{-\sqrt{\frac{T}{N}\sigma^2 + \frac{T^2}{N^2}\mu^2}} \ . \tag{12.13}$$

Substituting the expression for $\log(1 + y_{g,N})$ in (12.10) we readily get

$$p_N = \frac{1}{2} + \frac{1}{2}\frac{\mu}{\sigma}\frac{1}{\sqrt{\frac{N}{T} + (\frac{\mu}{\sigma})^2}} \ .$$

For N large, one can approximate $y_{g,N}, y_{b,N}$ and p_N by

$$y_{g,N} \quad = \quad e^{\sqrt{\frac{T}{N}}\sigma} - 1 \ ,$$

$$y_{b,N} \quad = \quad e^{-\sqrt{\frac{T}{N}}\sigma} - 1 \ ,$$

$$p_N \quad = \quad \frac{1}{2} + \frac{1}{2}\frac{\mu}{\sigma}\sqrt{\frac{T}{N}} \ .$$

In Chapter 14 we will study the limiting behavior of the stock price as N tends to infinity, i.e. as we subdivide the interval $[0, T]$ in increasingly smaller sub-intervals.

12.3   Equivalent Martingale Measures: Uniqueness

In this section we explore the existence of equivalent measures using the risk-free security as a numeraire. Recall that the discounted prices where given by

$$\tilde{B}_t \stackrel{\text{def}}{=} \frac{B_t}{B_t} \quad \text{and} \quad \tilde{S}_t \stackrel{\text{def}}{=} \frac{S_t}{B_t} = \frac{S_t}{(1+r)^t} .$$

Since $(\tilde{B}_t)$ is a martingale with respect to any probability measure, a probability measure Q on Ω is an equivalent martingale measure if and only if

$$0 < Q(\{\omega\}) < 1 \tag{12.14}$$

for all $\omega \in \Omega$ and the process $(\tilde{S}_t)$ is a Q-martingale, i.e.

$$\frac{S_t}{(1+r)^t} = E_Q[\frac{S_{t+1}}{(1+r)^{t+1}}|\mathcal{P}_t] \tag{12.15}$$

for all $t = 0, \ldots, N - 1$.

When is Q an equivalent martingale measure?

Before establishing the existence of an equivalent martingale measure we show that if such a measure exists it is unique and given by an explicit formula. We start by characterizing when a probability measure is an equivalent martingale measure. But before that we introduce the following notation which we recognize from the results in Chapter 2.

$$q \stackrel{\text{def}}{=} \frac{r - y_b}{y_g - y_b} . \tag{12.16}$$

Theorem 12.4 *Let Q be a probability measure. Then, Q is an equivalent martingale measure if and only if*

$$y_b < r < y_g \tag{12.17}$$

and

$$Q(Z_{t+1} = \delta|A) = q^\delta (1 - q)^{1-\delta} \tag{12.18}$$

for $t = 1, \ldots, N$, $A \in \mathcal{P}_t$, and $\delta \in \{0, 1\}$.
If Q is an equivalent martingale measure, then it is given by

$$Q(\{\omega\}) = q^{D_N(\omega)} (1 - q)^{N - D_N(\omega)} . \tag{12.19}$$

Proof Since S_t is $\mathcal{P}_t$-measurable and $1 + r$ is constant, condition (12.15) is equivalent to

$$1 + r = E_Q[\frac{S_{t+1}}{S_t}|\mathcal{P}_t] = E_Q[R_{t+1}|\mathcal{F}_t] . \tag{12.20}$$

Recall that by the definition of conditional expectation with respect to $\mathcal{P}_t$ we have

$$E_Q[R_{t+1}|\mathcal{P}_t] = \sum_{A \in \mathcal{P}_t} E_Q[R_{t+1}|A]\, 1_A .$$

Since $\mathcal{P}_t$ is a partition and the left-hand side of (12.20) is independent of $\omega \in \Omega$ we see that the martingale condition (12.15) is equivalent to

$$1 + r = E_Q[R_{t+1}|A] \tag{12.21}$$

for $A \in \mathcal{P}_t$. Set

$$q_A \overset{\text{def}}{=} Q(Z_{t+1} = 1|A) .$$

Inserting (12.4) into (12.21) using

$$Q(Z_{t+1} = 0|A) = 1 - q_A , \tag{12.22}$$

we readily verify that

$$\begin{aligned}
E_Q[R_{t+1}|A] &= E_Q[(1 + y_g)^{Z_{t+1}} (1 + y_b)^{1-Z_{t+1}}|A] \\
&= (1 + y_g)q_A + (1 + y_b) + 1 - q_A) .
\end{aligned}$$

Hence, (12.21) is equivalent to

$$1 + r = (1 + y_g)q_A + (1 + y_b)(1 - q_A) . \tag{12.23}$$

But (12.23) is true only if

$$q_A = \frac{r - y_b}{y_g - y_b} = q ,$$

as is easily seen by solving (12.23) for q_A . We have thus proved that (12.18) holds for all $A \in \mathcal{P}_t$.
We proceed to prove formula (12.19). Observe that for each $\omega = (\omega_1, \ldots, \omega_N) \in \Omega$ and $t = 1, \ldots, N$ the event

$$[Z_1 = \omega_1, \ldots, Z_t = \omega_t]$$

is an atom of the partition $\mathcal{P}_t$. Therefore, applying (12.18) and Proposition 4.11, we see that

$$\begin{aligned}
Q(\{\omega\}) &= Q(Z_1 = \omega_1, \ldots, Z_N = \omega_N) \\
&= Q(Z_1 = \omega_1)\, Q(Z_2 = \omega_2|Z_1 = \omega_1)\, Q(Z_3 = \omega_3|Z_1 = \omega_1, Z_2 = \omega_2) \\
&\quad \cdots Q(Z_N = \omega_N|Z_1 = \omega_1, Z_2 = \omega_2, \ldots, Z_{N-1} = \omega_{N-1}) \\
&= q^{\omega_1}(1 - q)^{1-\omega_1}\, q^{\omega_2}(1 - q)^{1-\omega_2} \cdots q^{\omega_N}(1 - q)^{1-\omega_N} \\
&= q^{\sum_{s=1}^{N} \omega_s}(1 - q)^{N - \sum_{s=1}^{N} \omega_s} \\
&= q^{D_N(\omega)}(1 - q)^{N - D_N(\omega)} .
\end{aligned}$$

This proves (12.19).

12.5 Pricing in the Cox–Ross–Rubinstein Economy

From now on we assume that

$$y_b < r < y_g$$

so that the market is complete and free of arbitrage opportunities. We thus know that any alternative $X : \Omega \to \mathbb{R}$ can be replicated by a self-financing strategy. At any time t the fair value of such an alternative is then given by the value of the replicating strategy. In Chapter 11 the latter was shown to be given by the $\mathcal{P}_t$-measurable random variable

$$\pi_t(X) = E_Q[X \frac{B_t}{B_N} | \mathcal{P}_t] = \frac{E_Q[X | \mathcal{P}_t]}{(1+r)^{N-t}} \qquad (12.24)$$

where Q is the equivalent martingale measure of the Cox–Ross–Rubinstein economy defined by (12.19).

Since $\pi_t(X)$ is $\mathcal{P}_t$-measurable, by Lemma 12.3, it will generally be a function of $S_0(\omega), S_1(\omega), \ldots, S_t(\omega)$. Stretching the notation slightly we can thus write

$$\pi_t(X; s_0, s_1, \ldots, s_t),$$

to denote the price at time t if the series of stock prices $S_0 = s_0, S_1 = s_1, \ldots, S_t = s_t$ has been observed.

At time $t = 0$ we can give a more explicit formula for $\pi_0(X)$:

$$\pi_0(X) = \frac{1}{(1+r)^N} \sum_{\omega \in \Omega} q^{D_N(\omega)} (1-q)^{N - D_N(\omega)} X(\omega), \qquad (12.25)$$

where q is given by

$$q = \frac{r - y_b}{y_g - y_b}.$$

We now investigate a class of alternatives for which it is possible to derive particularly simple pricing formulas and also having the property that $\pi_t(X; s_0, s_1, \ldots, s_t)$ is in fact only a function of s_t.

12.5.1 Valuing Path-Independent Alternatives

A *path-independent* alternative is a European alternative of the form

$$X_f = f(S_N),$$

where $f : \mathbb{R} \to \mathbb{R}$ is a given function. This class of alternatives is especially amenable because their payoff depends just on the final price and not on the particular path of the process (S_t). As we shall see in the next section European calls and puts belong to this class.

Since
$$S_N = S_0 \left(1 + y_g\right)^{D_N} \left(1 + y_b\right)^{N - D_N} ,$$

we find that
$$X_f = f(S_N) = h_f(S_0, D_N) , \tag{12.26}$$

where
$$h_f(x, l) \stackrel{\text{def}}{=} f(x(1 + y_g)^l (1 + y_b)^{N - l}) . \tag{12.27}$$

We start by giving the fair value formula at time $t = 0$.

Proposition 12.9 *At time* $t = 0$, *the fair value of an alternative of the form* (12.26) *is given by the formula*

$$\pi_0(X_f; S_0) = \frac{1}{(1 + r)^N} \sum_{l=0}^{N} \binom{N}{l} q^l (1 - q)^{N - l} h_f(S_0, l) ,$$

where h_f *is given by* (12.27).

Proof Using the representation of the pricing functional π_0 given by the equivalent martingale measure Q and carrying out some elementary manipulations, we find the desired formula

$$
\begin{aligned}
\pi_0(X_f) &= \frac{1}{(1 + r)^N} E_Q[X_f] = \frac{1}{(1 + r)^N} E_Q[h_f(S_0, D_N)] \\
&= \sum_{l=0}^{N} h_f(S_0, l) Q(D_t = l) = \sum_{l=0}^{N} \binom{N}{l} q^l (1 - q)^{N - l} h_f(S_0, l) ,
\end{aligned}
$$

where we have used that D_N is Bernoulli distributed with parameters N and q. $\square$

In order to derive an analogous formula for the value of X_f at an arbitrary time t in $\{0, \ldots, N\}$, we need the following lemma.

Lemma 12.10 *For an alternative of the form* (12.26) *we have*

$$E_Q[X_f | \mathcal{P}_t](\omega) = \sum_{j=0}^{N-t} \binom{N - t}{j} q^j (1 - q)^{N - t - j} h_f(S_t(\omega), j) , \tag{12.28}$$

for all $\omega \in \Omega$ *and* $t \in \{0, \ldots, N\}$. *Here* h_f *is given by* (12.27).

Proof We start by noting that

$$E_Q[X_f | \mathcal{P}_t] = \sum_{A \in \mathcal{P}_t} E_Q[X_f | A] 1_A . \tag{12.29}$$

Fix $\omega \in \Omega$ and let A be the unique atom in $\mathcal{P}_t$ containing ω. We will show that

$$E_Q[X_f|A] = \sum_{j=0}^{N-t} \binom{N-t}{j} q^j (1-q)^{N-t-j} h_f(S_t(\omega), j) \qquad (12.30)$$

holds. Since

$$E_Q[X_f|\mathcal{P}_t](\omega) = E_Q[X_f|A]$$

this will prove the lemma.

It remains to prove (12.30). Letting

$$l \stackrel{\text{def}}{=} D_t(\omega)$$

we see that

$$S_t(\omega) = S_0(1+y_g)^l(1+y_b)^{t-l} \qquad (12.31)$$

holds. Moreover, $D_t = l$ on A.

First note that $Q([D_N = k]|A) = 0$ whenever $k < l$ or $k > N - (t - l)$. For $l \le k \le N - (t-l)$ we have

$$
\begin{aligned}
Q([D_N = k]|A) &= \frac{Q([D_N = k] \cap A)}{Q(A)} = \frac{Q([\sum_{s=t+1}^{N} Z_s = k - l] \cap A)}{Q(A)} \\
&= \frac{Q([\sum_{s=t+1}^{N} Z_s = k - l])\, Q(A)}{Q(A)} \\
&= Q([\sum_{s=t+1}^{N} Z_s = k - l]) \\
&= \binom{N-t}{k-l} q^{k-l}(1-q)^{N-t-(k-l)} ,
\end{aligned}
$$

where we have used the independence of the two events $[\sum_{s=t+1}^{N} Z_s = k-l]$ and A, as well as the fact that $\sum_{s=t+1}^{N} Z_s$ counts the number of successes when repeating $N - t$ times a Bernoulli game with probability of success q.

Using this we see that

$$
\begin{aligned}
E_Q[X_f|A] &= E_Q[h_f(S_0, D_N)|A] \\
&= \sum_{k=0}^{N} Q(D_N = k|A) h_f(S_0, k) \\
&= \sum_{k=l}^{N-t-l} \binom{N-t}{k-l} q^{k-l}(1-q)^{N-t-(k-l)} h_f(S_0, k) \\
&= \sum_{j=0}^{N-t} \binom{N-t}{j} q^j (1-q)^{N-t-j} h_f(S_0, l+j)
\end{aligned}
$$

where the last equality was obtained by substitution j for $k - l$. Using that

$$h_f(S_0, l+j) = h_f(S_t(\omega), j)$$

holds, we obtain(12.30), concluding the proof of the lemma. □

Proposition 12.11 *The price of $\pi_t(X_f; S_0, S_1(\omega), \ldots, S_t(\omega))$ of X_f at time t depends only on $S_t(\omega)$ and is given by the formula*

$$\pi_t(X_f; S_t) = \frac{1}{(1+r)^{N-t}} \sum_{l=0}^{N-t} \binom{N-t}{l} q^l (1-q)^{N-t-l} h_f(S_t(\omega), l) \,,$$

where h_f is given by (12.27).

Proof The result follows immediately from the lemma and the fact that

$$\pi_t(X_f) = \frac{1}{(1+r)^{N-t}} E_Q[X_f | \mathcal{P}_t] \,.$$

□

12.6   Hedging in the Cox–Ross–Rubinstein Economy

Let $X : \Omega \to \mathbb{R}$ be an alternative. Since the Cox–Ross–Rubinstein economy is complete we find a replicating strategy $\Phi = (\Phi_t)$ for X. In the Cox–Ross–Rubinstein economy the portfolio Φ_t is given by a pair (ϕ_t^0, ϕ_t^1), where ϕ_t^0 denotes the number of units of the risk-free security and ϕ_t^1 the number of units of the stock. Recall from Chapter 9 that Φ being a replicating portfolio for X means that

- Φ is predictable, i.e. Φ_t is $\mathcal{P}_{t-1}$-measurable;

- Φ is self-financing, i.e.

$$\phi_t^0 B_{t+1} + \phi_t^1 S_{t+1} = V_t(\Phi) = \phi_{t+1}^0 B_{t+1} + \phi_{t+1}^1 S_{t+1} \,;$$

  and

- $V_N(\Phi) = \phi_N^0 B_N + \phi_N^1 S_N = X \,.$

Since (Φ_t) is predictable, by Lemma 12.3, Φ_t will be a function of the random variables $S_0, S_1, \ldots, S_{t-1}$. As in the previous section, by a slight abuse of notation, we may write

$$\phi_t^0 = \phi_t^j(s_0, s_1, \ldots, s_{t-1}) \quad \text{and} \quad \phi_t^1 = \phi_t^j(s_0, s_1, \ldots, s_{t-1}),$$

to denote, respectively, the number of units acquired at time t of the money market account and the stock if up to time t the series of stock prices $S_0 = s_0, S_1 = s_1, \ldots, S_t = s_t$ has been observed. We will now try to describe Φ_{t+1} in terms of the possible values of X at time $t + 1$.

By definition of the value of X we have

$$\pi_{t+1}(X; S_0, S_1, \ldots, S_{t+1}) = V_{t+1}(\Phi) = \phi_t^0 B_{t+1} + \phi_t^1 S_{t+1}.$$

Since S_{t+1} can only have the values $S_t(1 + y_g)$ or $S_t(1 + y_b)$, the above equality between random variables translates into the following two linear equations

$$\phi_{t+1}^0 (1 + r)^{t+1} + \phi_{t+1}^1 S_t(1 + y_g) = \pi_{t+1}(X; S_0, S_1, \ldots, S_t(1 + y_g)),$$
$$\phi_{t+1}^0 (1 + r)^{t+1} + \phi_{t+1}^1 S_t(1 + y_b) = \pi_{t+1}(X; S_0, S_1, \ldots, S_t(1 + y_b)).$$

Solving these equations for ϕ_t^0 and ϕ_t^1 gives

$$\phi_{t+1}^1 = \frac{\pi_{t+1}(X; S_0, \ldots, S_t(1 + y_g)) - \pi_{t+1}(X; S_0, \ldots, S_t(1 + y_b))}{S_t(y_g - y_b)}.$$

$$\phi_{t+1}^0 = \frac{\pi_{t+1}(X; S_0, \ldots, S_t(1 + y_b))(1 + y_g) - \pi_{t+1}(X; S_0, \ldots, S_t(1 + y_g))(1 + y_b)}{(1 + r)^{t+1}(y_g - y_b)}.$$

It is not difficult to see the way this relates to the one-period model studied in Chapter 2. In fact the replicating portfolio Φ_t corresponds to replicating in a one-period context the claim which pays $\pi_{t+1}(X; S_0, S_1, \ldots, S_t(1 + y_g))$ and $\pi_{t+1}(X; S_0, S_1, \ldots, S_t(1 + y_b))$ if the state of the world is "good" or "bad", respectively. Thus, replication in the multi-period context turns out to be nothing but a series of replications in one-period sub-models.

12.7 European Call and Put Options

We are particularly interested in two contingent claims which play an important role in the financial markets: European call and put options. A *European call option* on the stock gives the holder the right but not the obligation to buy at a given time (the *maturity* of the option) one unit of the stock at a pre-specified price (the *strike or exercise price* of the option). In monetary terms, owning a European call option is equivalent to holding the claim with payoff

$$X_{\text{call}}(\omega) = \max\{ S_N(\omega) - K , 0 \}, \quad \omega \in \Omega = \{0, 1\}^N,$$

where K is the strike price and N is the maturity of the option.

Analogously a *European put option* on the stock gives the holder the right but not the obligation to sell at a given time one unit of the stock at a pre-specified price. The monetary equivalent of a European put option is the claim with payoff

$$X_{\mathrm{put}}(\omega) = \max\{\, K - S_N(\omega)\,,\, 0\,\}\,, \quad \omega \in \Omega = \{0,1\}^N\,,$$

where again K is the strike price and N is the maturity of the option.
Before giving pricing formulas for put and calls we look at an important relationship between their prices.

Put-call parity

The put-call parity is based on the simple observation that buying a call and selling a put with the same strike price K and maturity N has the same net payoff as the European claim

$$X = S_N - K\,.$$

Proposition 12.12 *The prices of European call and put options with the same maturity and the same strike price K are related by*

$$\pi_0[\,X_{\mathrm{call}}\,] - \pi_0[\,X_{\mathrm{put}}\,] = S_0 - \frac{K}{(1+r)^N}\,.$$

Proof Since

$$X_{\mathrm{call}} - X_{\mathrm{put}} = \max\{\, S_N - K\,,\, 0\,\} - \max\{\, K - S_N\,,\, 0\,\} = S_N - K$$

holds, the linearity of the pricing functional implies

$$\pi_0[\,X_{\mathrm{call}}\,] - \pi_0[\,X_{\mathrm{put}}\,] = \pi_0[\,X_{\mathrm{call}} - X_{\mathrm{put}}\,] = \pi_0[\,S_N - K\,] = \pi_0[S_N] - \pi_0[K]\,.$$

Since we know that π_0 reproduces the prices of the traded securities, i.e.

$$\pi_0[S_N] = S_0 \ \ \text{and} \ \ \pi_0[K] = \frac{K}{(1+r)^N}\,,$$

we immediately obtain the desired formula

$$\pi_0[\,X_{\mathrm{call}}\,] - \pi_0[\,X_{\mathrm{put}}\,] = S_0 - \frac{K}{(1+r)^N}\,.$$

□

We note that the put-call parity is just a consequence of the linearity of the pricing functional π_0 and does not depend on any other model assumptions on the economy. In particular, the put-call parity permits us to derive the price of a put when the price of the corresponding call is known.

Pricing Formula

In order to formulate the pricing formula in a more transparent way we introduce
the following notation.

$$m(K) \stackrel{def}{=} \inf_{n \in \mathbb{N}} \{ \, n > \log[\frac{K}{S_0(1+y_b)^N}] \, \Big/ \, \log[\frac{1+y_g}{1+y_b}] \, \} \, .$$

Thus, $m(K)$ is the least integer greater than $\log[\frac{K}{S_0(1+y_b)^N}] \Big/ \log[\frac{1+y_g}{1+y_b}]$. A few
algebraic manipulations show that the following holds.

Lemma 12.13

$$\max\{ \, S_N(\omega) - K \, , \, 0 \, \} > 0 \Leftrightarrow D_N(\omega) \geq m(K) \, .$$

Using this equivalence, a direct application of Proposition 12.9 to X_{call} yields

$$\pi_0(X_{\text{call}}; S_0) \quad = \quad \frac{S_0}{(1+r)^N} \sum_{l=m(K)}^{N} \binom{N}{l} q^l (1-q)^{N-l} (1+y_g)^l (1+y_b)^{N-l}$$

$$- \frac{K}{(1+r)^N} \sum_{l=m(K)}^{N} \binom{N}{l} q^l (1-q)^{N-l} \, .$$

In order to simplify this formula we define for any $0 < p < 1$,

$$G(k, N, p) \stackrel{def}{=} \sum_{j=k}^{N} \binom{n}{j} p^j (1-p)^{N-j} \, .$$

Thus, by the results of Chapter 8.4, $G(k, N, p) = P(k \leq X \leq N)$ for a Bernoulli
distributed random variable X with parameters p and N. With this notation and
after a few algebraic manipulations, the above formula for the value of a European
call can be re-stated as follows.

Proposition 12.14

$$\pi_0(X_{call}, S_0) = S_0 \, G(m(K), N, q^*) - \frac{K}{(1+r)^N} \, G(m(K), N, q) \, ,$$

where $0 < q^* < 1$ is defined by

$$q^* \stackrel{def}{=} q \, \frac{1+y_g}{1+r} \, .$$

Proof The result is easily obtained by straightforward algebraic manipulations.
We only note that with the above definition of q^* one easily shows that $0 < q^* < 1$
and

$$1 - q^* = (1-q)\frac{1+y_b}{1+r} \, .$$

$\square$

Concluding Remarks and Suggestions for Further Reading

The Cox–Ross–Rubinstein model is a straightforward generalization of the simple model of Chapter 2. This simple model has played a major role in applications since it is particularly easy to implement and intuitive to understand. The article [17] represented a major step in the study of option pricing among other things because it also showed that the celebrated Black–Scholes formula could be obtained as a limiting case by letting the length of the time intervals tend to zero. The Black–Scholes formula had up to then only been accessible to people with a command of the technically demanding stochastic calculus. To prepare the ground for the study of the limiting behavior of the Cox–Ross–Rubinstein model we present in the next chapter a special case of the Central Limit Theorem. An exhaustive treatment of the Cox–Ross–Rubinstein model is given in [18].

Proof That F is increasing follows immediately by Proposition 13.2 and the fact that if $x \leq y$ we have that $(-\infty, x] \subset (-\infty, y]$.

Take now any decreasing sequence (x_n) converging to x_0 and set $A_n \stackrel{\text{def}}{=} (-\infty, x_n]$. Then (A_n) is a decreasing sequence with $(-\infty, x_0] = \cap_{n=1}^{\infty} A_n$. By Proposition 13.3 we obtain

$$\lim_{n \to \infty} F(x_n) = \lim_{n \to \infty} P(A_n) = P((-\infty, x_0]) = F(x_0)$$

Let now $A_N \stackrel{\text{def}}{=} (-\infty, N]$. Then (A_N) is an increasing sequence with $\Omega = \cup_{N=1}^{\infty} A_N$. Thus, by Proposition 13.3 we obtain

$$\lim_{N \to \infty} F(N) = \lim_{N \to \infty} P(A_N) = P(\Omega) = 1 .$$

For every $\varepsilon > 0$ we can thus choose $N(\varepsilon)$ such that

$$1 - \varepsilon \leq F(N) \leq 1$$

holds for all $N \geq N(\varepsilon)$. Since F is increasing we obtain

$$1 - \varepsilon \leq F(x) \leq 1$$

for all $x \geq N(\varepsilon)$. Thus, $\lim_{x \to \infty} F(x) = 1$.

That $\lim_{x \to -\infty} F(x) = 0$ holds is proved analogously. $\qquad \square$

Remark 13.7 *The points at which a distribution function is continuous are easily identified as the points $x_0 \in \mathbb{R}$ for which $P(X = x_0) = 0$. Indeed, let (x_n) be an increasing sequence converging to $x_0 \in \mathbb{R}$. Setting $A_n \stackrel{\text{def}}{=} (-\infty, x_n]$ we have that (A_n) is an increasing sequence with $(-\infty, x_0) = \cup_{n=1}^{\infty} A_n$. From Proposition 13.3 we get that*

$$\begin{aligned}
\lim_{n \to \infty} F(x_n) &= \lim_{n \to \infty} P(A_n) = P((-\infty, x_0)) \\
&= P((-\infty, x_0]) - P(\{x_0\}) \\
&= F(x_0) - P(\{x_0\}) .
\end{aligned}$$

Therefore, F is left-continuous in x_0 if and only if $P(\{x_0\}) = 0$. Since we have already proved that F is always right-continuous, it follows that F is continuous in x_0 if and only if $P(\{x_0\}) = 0$.

The distribution function of a random variable can have at most countable many points of discontinuity. This follows from a well-known fact of elementary analysis (see for instance [50]) stating that every monotone function can have at most a countable number of points of discontinuity.

If F is the distribution function of P, then

$$P((a, b]) = P((-\infty, b]) - P((-\infty, a]) = F(b) - F(a) .$$

In fact this is the key to constructing probability measures on $(\mathbb{R}, \mathcal{B}(\mathbb{R}))$. Given any increasing, right-continuous function $F : \mathbb{R} \to [0, 1]$ satisfying (13.1) we can set

$$P((a, b]) \stackrel{\text{def}}{=} F(b) - F(a)$$

Although this function is defined only on the collection of intervals of the form $(a, b]$ it can be extended in a unique way to a probability measure on $(\mathbb{R}, \mathcal{B}(\mathbb{R}))$. This is the only non-elementary result on general probability spaces we will use without proof. The proof is in itself not difficult but would divert us from our main objective. It can be found in all textbooks on a measure theoretic approach to probability (see for instance [3], [8], [36] or [55]).

Theorem 13.8 *Given any increasing, right-continuous function $F : \mathbb{R} \to [0, 1]$ satisfying (13.1) there exists a unique probability measure P on $(\mathbb{R}, \mathcal{B}(\mathbb{R}))$ which has F as a distribution function.*

Because of this result any increasing, right-continuous function F satisfying (13.1) is commonly called a *distribution function*.

Probabilities with continuous density functions

An important class of distribution function consists of those distribution functions which are given by a density function. We will only consider continuous density functions in this book.

A (continuous) *density function* is a continuous function $f : \mathbb{R} \to \mathbb{R}$ which is non-negative and satisfies

$$\int_{-\infty}^{\infty} f(x)dx = 1 .$$

Here, the integrals are understood in the Riemann sense (see [50]. Defining

$$F(a) = \int_{-\infty}^{a} f(x)dx ,$$

it is easy to see from the properties of the integral that F is an increasing, continuous (in fact continuously differentiable) function satisfying (13.1). Thus, it is a distribution function. Therefore, by Theorem 13.8, every density function defines a probability measure $P : \mathcal{B} \to [0, 1]$ on $(\mathbb{R}, \mathcal{B}(\mathbb{R}))$.

Remark 13.9 *If P is given by a density function, then*

$$P((a, b]) = F(b) - F(a) = \int_{a}^{b} f(x)dx .$$

In particular, $P(\{a\}) = 0$ for any $a \in \mathbb{R}$, which by Remark 13.7 again shows that the distribution function of P is continuous.

An important example of a probability function given by a density function is the normal distribution described next. The following result can be found in most books on calculus.

Lemma 13.10 *Define*

$$\varphi_{\mu,\sigma}(x) \stackrel{def}{=} \frac{1}{\sqrt{2\pi}\sigma} e^{-\frac{(x-\mu)^2}{2\sigma^2}} .$$

Then,

$$\int_{-\infty}^{\infty} \varphi_{\mu,\sigma}(x)dx = 1.$$

It follows that $\varphi_{\mu,\sigma}$ is a continuous density function with corresponding distribution function

$$F_{\mathcal{N}_{\nu,\sigma}}(x) = \int_{-\infty}^{x} \varphi_{\mu,\sigma}(x)dx . \tag{13.2}$$

The function $F_{\mathcal{N}_{\nu,\sigma}}$ is commonly called the *normal distribution function* with parameters μ and σ.

13.3   Random Variables

Let $(\Omega, \mathcal{A}, P)$ be an arbitrary probability space. A function $X : \Omega \to \mathbb{R}$ is said to be a random variable if for each Borel set $B \in \mathcal{B}(\mathbb{R})$ we have

$$X^{-1}(B) \in \mathcal{A} . \tag{13.3}$$

Compare this with our definition of measurability with respect to an algebra in Section 7.1.3. The reason for singling out those functions $X : \Omega \to \mathbb{R}$ which have the above property becomes apparent when we think that we would like to assess the probability of the random variable taking values in a given Borel set B, i.e. we want to know

$$P([X \in B]) .$$

For the above expression to make sense we have to require (13.3).
A random variable is said to be *discrete* if it can only assume finitely or denumerably many values $x_1, x_2, \dots$. Otherwise it is said to be a *continuous* random variable.

Lemma 13.11 *Let $X : \Omega \to \mathbb{R}$ be a function. Then X is a random variable if and only if*

$$X^{-1}((a,b)) \in \mathcal{A}$$

holds for all open intervals (a, b).

Proof We need only prove the sufficiency of the condition. Set

$$\mathcal{B}_X \stackrel{\text{def}}{=} \{B \subset \mathbb{R}; X^{-1}(B) \in \mathcal{A}\}.$$

Then $\mathcal{B}_X$ is a σ-algebra. Indeed, if $B \in \mathcal{B}_X$ we have that $X^{-1}(B^c) = X^{-1}(B)^c \in \mathcal{A}$ and, hence, $B^c \in \mathcal{B}_X$. If (B_i) is a sequence in $\mathcal{B}_X$, we have that $X^{-1}(\cup_i B_i) = \cup_i X^{-1}(B_i) \in \mathcal{A}$. Thus, $\cup_i B_i$ belongs to $\mathcal{B}_X$. Finally, $X^{-1}(\mathbb{R}) = \Omega \in \mathcal{A}$, so that $\mathbb{R} \in \mathcal{B}_X$.

Since obviously $\mathcal{I}_{\text{open}} \subset \mathcal{B}_X$, we have that $\mathcal{B}(\mathbb{R}) = \mathcal{A}(\mathcal{I}_{\text{open}}) \subset \mathcal{B}_X$. Thus, X is measurable. $\square$

The following establishes that continuous functions of a random variable are again random variables. In fact the result holds for classes of functions more general than the class of continuous functions, but we will not need this in the sequel.

Lemma 13.12 *Let $f : \mathbb{R} \to \mathbb{R}$ be continuous and assume $X : \Omega \to \mathbb{R}$ is a random variable. Then $f(X)$ is also a random variable.*

Proof Set $Y \stackrel{\text{def}}{=} f(X)$. Take an open interval I. Since f is continuous, $f^{-1}(I)$ is open and can be written as a countable union $\cup_i I_i$ of open intervals I_i. Therefore,

$$Y^{-1}(I) = \cup X^{-1} f^{-1}(I_i) \in \mathcal{A}$$

since X is measurable. This completes the proof. $\square$

Distribution of a random variable

Let $X : \Omega \to \mathbb{R}$ be a random variable and set

$$P_X(B) \stackrel{\text{def}}{=} P([X \in B]).$$

Then P_X is a probability measure on $(\mathbb{R}, \mathcal{B}(\mathbb{R}))$ which is called the *distribution* of X.

Expected value of a random variable

Assume $(\Omega, \mathcal{A}, P)$ to be an arbitrary probability space and $X : \Omega \to \mathbb{R}$ a discrete random variable with range $x_1, x_2, \ldots$.

For such a random variable X it makes sense to define its expected value by

$$E_P[X] \stackrel{\text{def}}{=} \sum_{i=1}^{\infty} x_i P(X = x_i)$$

under the condition that the above series converges and its value is independent of a particular ordering. This is the case if and only if the series converges absolutely, i.e. if

$$\sum_{i=1}^{\infty} |x_i| P(X = x_i) < \infty .$$

Denote by $\mathcal{D}_1(\Omega, P)$ the set of discrete random variables for which $E_P[X]$ is well defined.

Lemma 13.13 *Let* $X, Y \in \mathcal{D}_1(\Omega, P)$ *and* $\lambda \in \mathbb{R}$. *Then:*

a) $E_P[1_\Omega] = 1$.

b) $X + \lambda Y \in \mathcal{D}_1(\Omega, P)$ *and*

$$E_P[X + \lambda Y] = E_P[X] + \lambda E_P[Y] .$$

c) *If* $X \leq Y$ *holds, then* $E_P[X] \leq E_P[Y]$.

d) $E_P[X] \leq \sup_{\omega \in \Omega} X(\omega)$.

e) $|E_P[X]| \leq E_P[|X|]$.

Proof Property a) is immediate from the definition.
To prove b) let

$$X(\Omega) = \{x_1, x_2, x_3, \ldots\} \quad \text{and} \quad Y(\Omega) = \{y_1, y_2, y_3, \ldots\} .$$

Then:

$$
\begin{aligned}
E_P[X + \lambda Y] &= \sum_i \sum_j (x_i + \lambda y_j) P([X = x_i] \cap [Y = y_j]) \\
&= \sum_i x_i \sum_j P([X = x_i] \cap [Y = y_j]) \\
&\quad + \lambda \sum_j y_j \sum_i P([X = x_i] \cap [Y = y_j]) \\
&= \sum_i x_i P([X = x_i]) + \lambda \sum_j y_j P([Y = y_j]) \\
&= E_P[X] + \lambda E_P[Y] .
\end{aligned}
$$

To prove c) first let $X \geq 0$ hold. Then, $x_i \geq 0$ for all i. Then it is obvious from the definition that $0 \leq 0 E_P[X]$. Now, if $X \leq Y$ holds, then $0 \leq Y - X$. Hence, using the linearity of $E_P[\cdot]$ we obtain

$$0 \leq E_P[Y - X] = E_P[Y] - E_P[X] .$$

as claimed.

To see that d) holds just note that

$$X \leq \sup_{w \in \Omega} X(w) .$$

Thus, we get from c) that

$$E_P[X] \leq E_P[\sup_{w \in \Omega} X(w)] = \sup_{w \in \Omega} X(w) .$$

$\square$

Let now $X : \Omega \to \mathbb{R}$ be an arbitrary random variable. We will define $E_P[X]$ by approximating X by a suitable sequence of discrete random variables X_n. Observe that for each natural number n we can write Ω as the disjoint union

$$\Omega = \bigcup_{k=-\infty}^{\infty} [\frac{k}{n} \leq X < \frac{k+1}{n}] .$$

Define now

$$X_n \stackrel{\text{def}}{=} \sum_{k=-\infty}^{\infty} \frac{k}{n} 1_{[\frac{k}{n} \leq X < \frac{k+1}{n}]} .$$

Lemma 13.14   a) For all n and m we have

$$X_m \leq X < X_n + \frac{1}{n} \tag{13.4}$$

b)

$$|X - X_n| \leq \frac{1}{n} . \tag{13.5}$$

c) If $X_n \in \mathcal{D}_1(\Omega, P)$ for some n, then the same is true for all other n.

d) If $X_1 \in \mathcal{D}_1(\Omega, P)$, then $(E_P[X_n])_{1 \leq n}$ is a Cauchy sequence.

Proof Inequality (13.4) is obtained directly from the definition of X_m and X_n. Choosing $n = m$ in (13.4), it follows that

$$0 \leq X - X_n \leq \frac{1}{n} ,$$

which implies (13.5) and the uniform convergence. From (13.4) we also infer that

$$X_n - X_m \leq \frac{1}{m} \quad \text{and} \quad X_m - X_n \leq \frac{1}{n} ,$$

and, therefore,

$$|X_n - X_m| \leq \frac{1}{n} + \frac{1}{m} . \tag{13.6}$$

This implies that

$$|X_m| \leq |X_n| + \frac{1}{n} + \frac{1}{m} .$$

From this we conclude that if $X_n \in \mathcal{D}_1(\Omega, P)$ for some n, so must be X_m. Assume now that $X_1 \in \mathcal{D}_1(\Omega, P)$, so that $E_P[X_n]$ exists for all n. Then, from Lemma 13.13 and (13.6) we obtain

$$|E_P[X_n] - E_P[X_m]| \leq E_P[|X_n - X_m|] \leq \frac{1}{n} + \frac{1}{m} ,$$

proving that $(E_P[X_n])_{1 \leq n}$ is a Cauchy sequence.   □

Assume now that $X_1 \in \mathcal{D}_1(\Omega, P)$. Since $(E_P[X_n])_{1 \leq n}$ is a Cauchy sequence, we can define

$$E_P[X] \stackrel{\text{def}}{=} \lim_{n \to \infty} E_P[X_n] .$$

Remark 13.15 *We still have to check whether for discrete random variables this definition is consistent with our old definition. To see this let X be discrete and (X_n) the approximating sequence. From Lemma 13.13 and (13.5) we conclude that*

$$|E_P[X] - E_P[X_n]| \leq E_P[|X - X_n|] \leq \frac{1}{n} ,$$

where $E_P[\cdot]$ denotes the expectation operator for discrete random variables. This implies that the limit of $E_P[X_n]$ coincides with our old definition of $E_P[X]$ for discrete random variables.

Random variables X for which $E_P[X]$ exists are called integrable. The set of integrable random variables will be denoted by $\mathcal{L}_1(\Omega, P)$.

Proposition 13.16 *Let $X, Y \in \mathcal{L}_1(\Omega, P)$ and $\lambda \in \mathbb{R}$. Then*

   a) $X + \lambda Y \in \mathcal{L}_1(\Omega, P)$ and $E_P[X + \lambda Y] = E_P[X] + \lambda E_P[Y]$.

   b) If $X \leq Y$ holds, then $E_P[X] \leq E_P[Y]$.

   c) $E_P[X] \leq \sup_{\omega \in \Omega} X(\omega)$.

   d) $|E_P[X]| \leq E_P[|X|]$.

Proof We first prove a). Let $X, Y \in \mathcal{L}_1(\Omega, P)$ and $\lambda \in \mathbb{R}$ be given. We denote by X_n and Y_n their respective approximations and by $[X + \lambda Y]$ the approximation of $X + \lambda Y$. Then, repeated use of (13.4) on X, Y and $X + \lambda Y$ yields

$$\begin{aligned}
[X + \lambda Y]_n &\leq X + \lambda Y \leq X_n + \lambda Y_n + \frac{2}{n} \\
&\leq X + \lambda Y + \frac{2}{n} \leq [X + \lambda Y]_n + \frac{3}{n} .
\end{aligned}$$

From this we first read that if $E_P[X]$ and $E_P[Y]$ exist, so does $E_P[X + \lambda Y]$. We also obtain

$$E_P[[X + \lambda Y]_n] \le E_P[X_n] + \lambda E_P[Y_n] \le E_P[[X + \lambda Y]_n] + \frac{3}{n} .$$

Taking the limit as n tends to infinity we obtain

$$E_P[X + \lambda Y] \le E_P[X] + \lambda E_P[Y] \le E_P[X + \lambda Y] ,$$

and thus $E_P[X + \lambda Y] = E_P[X] + \lambda E_P[Y]$.
To prove b) we again invoke Lemma 13.14 to see that

$$X_n \le X \le Y \le Y_n + \frac{1}{n} .$$

This gives

$$E_P[X_n] \le E_P[Y_n] + \frac{1}{n} ,$$

which in turn implies $E_P[X] \le E_P[Y]$. The last assertion is proved in the same way as in Lemma 13.13. $\qquad\square$

Theorem 13.17 *Let $X : \Omega \to \mathbb{R}$ be a random variable such that its distribution P_X has a continuous density function $f : \mathbb{R} \to \mathbb{R}$. Then for any continuous $g : \mathbb{R} \to \mathbb{R}$ we have*

$$E_P[g(X)] = \int_{-\infty}^{\infty} g(x)f(x)dx ,$$

where the integral on the right is a Riemann integral.

Proof For each $n \ge 1$ and $-\infty \le k \le \infty$ we have the inequalities

$$\frac{k}{n} \int_{\frac{k}{n}}^{\frac{k+1}{n}} f(x)dx \le \int_{\frac{k}{n}}^{\frac{k+1}{n}} xf(x)dx \le \frac{k+1}{n} \int_{\frac{k}{n}}^{\frac{k+1}{n}} f(x)dx . \qquad (13.7)$$

Recalling that the approximating function X_n used the definition of $E_P[X]$ and the fact that

$$P(\frac{k}{n} < X \le \frac{k+1}{n}) = \int_{\frac{k}{n}}^{\frac{k+1}{n}} f(x)dx ,$$

we obtain from (13.7) by summing over all $-\infty \le k \le \infty$,

$$
\begin{aligned}
E_P[X_n] &= \sum_{-\infty}^{\infty} \frac{k}{n} P(\frac{k+1}{n} < X \le \frac{k+1}{n}) \\
&\le \int_{-\infty}^{\infty} xf(x)dx \\
&\le \sum_{-\infty}^{\infty} \frac{k}{n} P(\frac{k+1}{n} < X \le \frac{k+1}{n}) + \frac{1}{n} = E_P[X_n] + \frac{1}{n} .
\end{aligned}
$$

Letting n tend to infinity we obtain

$$E_P[X] = \int_{-\infty}^{\infty} x f(x) dx .$$

Inequality (13.7) also shows that $X \in \mathcal{L}_1(\Omega, P)$ if and only if $\int_{-\infty}^{\infty} |x| f(x) dx$ exists in the Riemann sense.

For each N choose a sequence (x_k^N) in $\mathbb{R}$ such that

$$\lim_{k \to -\infty} x_k^N = -\infty \quad \text{and} \quad \lim_{k \to \infty} x_k^N = \infty ,$$

and

$$g(x_k^N) - \frac{1}{N} \leq g(x) \leq g(x_k^N) + \frac{1}{N}$$

for $x_k^N \leq x_{k+1}^N$. Set

$$g_N(x) \stackrel{\text{def}}{=} x_k^N \qquad \text{if } x_k^N \leq x_{k+1}^N .$$

Note that by definition

$$|g_N(x) - g(x)| \leq \frac{1}{N} ,$$

for all $x \in \mathbb{R}$ so that

$$|E_P[g_N(X)] - E_P[g(X)]| \leq E_P[|g_N(X) - g(X)|] \leq \frac{1}{N} .$$

It follows that

$$\lim_{N \to \infty} E_P[g_N(X)] = E_P[g(X)] .$$

By definition of g_N we have the inequalities

$$(g_N(x_k^N) - \frac{1}{N}) \int_{x_k^N}^{x_{k+1}^N} f(x) dx \leq \int_{x_k^N}^{x_{k+1}^N} g(x) f(x) dx \leq (g_N(x_k^N) + \frac{1}{N}) \int_{x_k^N}^{x_{k+1}^N} f(x) dx .$$

Noting that $P(x_k^N \leq X < x_{k+1}^N) = \int_{x_k^N}^{x_{k+1}^N} f(x) dx$ and summing over $-\infty < k < \infty$ we find

$$E_P[g_N(X)] - \frac{1}{N} \leq \int_{x_k^N}^{x_{k+1}^N} g(x) f(x) dx \leq E_P[g_N(X)] + \frac{1}{N} .$$

Letting N tend to infinity we obtain

$$E_P[g(X)] = \int_{-\infty}^{\infty} g(x) f(x) dx .$$

$\square$

Corollary 13.18 *If for $k \in \mathbb{N}$ we have that $X^k \in \mathcal{L}(\Omega, P)$, then*

$$E_P[X^k] = \int_{-\infty}^{\infty} x^k f(x) dx$$

If $X^2 \in \mathcal{L}(\Omega, P)$, define its variance as

$$\text{Var}_P[X] \stackrel{\text{def}}{=} E_P[(X - E_P[X])^2] = E_P[X]^2 - E_P[X^2] .$$

The normal distribution

A random variable is said to be *normally distributed* with parameters μ and σ if its distribution function is equal to $F_{\mathcal{N}_{\nu,\sigma}}$ defined by (13.2), i.e.

$$F_{\mathcal{N}_{\nu,\sigma}}(x) = \frac{1}{\sqrt{2\pi}\sigma} \int_{-\infty}^{x} e^{-\frac{1}{2}\left(\frac{t-\mu}{\sigma}\right)^2} dt .$$

The function $F_{\mathcal{N}_{\nu,\sigma}}$ is sometimes also-called the cumulative normal distribution. The following result clarifies the role of the parameters μ and σ.

Theorem 13.19 *Let $X : \Omega \to \mathbb{R}$ be a normally distributed random variable with parameters μ and σ. Then*

$$E_P[X] = \mu \qquad \text{and} \qquad \text{Var}_P[X] = \sigma^2 .$$

Bernoulli or binomially distributed random variables

Let B be a binomially distributed random variable with parameters N and $p \in (0, 1)$. Recall that it's distribution function is given by

$$F_B(x) = \sum_{j=0}^{\lfloor x \rfloor} \binom{N}{j} p^j (1 - p)^{N-j} ,$$

where

$$\lfloor x \rfloor := \max_{z \in \mathbb{Z}} \{ z \leq x \} .$$

Observe that $F_B(x) = 0$ for $x < 0$ and that $F_B(x) = 1$ for $x \geq N$. Note that the distribution function $F_{\mathcal{N}}(\cdot)$ is continuous whereas $F_B(\cdot)$ is not continuous.

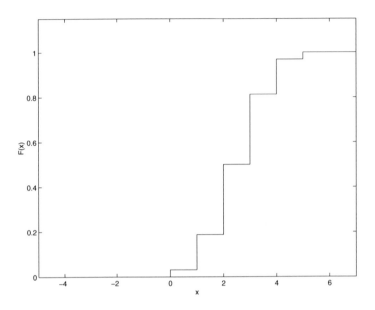

Figure 13.2: Distribution Function of a Binomial Random Variable for $N = 5$ and $p = 0.5$.

13.4   Weak Convergence of a Sequence of Random Variables

Definition 13.20 *A sequence of random variables* (X_n), $n \geq 1$, *is said to converge weakly, or in distribution, towards a random variable X if*

$$\lim_{n \to \infty} F_{X_n}(s) = F_X(s)$$

at each point $s \in \mathbb{R}$ where the distribution function $F_X(\cdot)$ is continuous. In that case we often use the notation

$$X_n \overset{w}{\to} X .$$

Recall that in Remark 13.7 the points of continuity were characterized as those $a \in \mathbb{R}$ for which $P(X = a) = 0$. Moreover, there are at most countable many of them. We will use this fact below.

We next give a slightly more intuitive formulation of weak convergence.

Lemma 13.21 *A sequence of random variables* (X_n), $n \geq 1$, *converges weakly towards a random variable* X *if and only if for all pairs of numbers* a, b *with* $a < b$ *and*

$$P(X = a) = P(X = b) = 0 \, ,$$

we have

$$\lim_{n \to \infty} P([a < X_n \leq b]) = P([a < X \leq b]) \, .$$

Since in applications often the convergence is towards a normally distributed random variable the following lemma is useful.

Lemma 13.22 *If a sequence of random variables* (X_n), $n \geq 1$, *converges weakly towards a normal random variable* N, *then their distribution functions* F_{X_n} *converge to* F_N *uniformly on compact subsets of* $\mathbb{R}$.

Proof We first show local uniform convergence, i.e. to any $s \in \mathbb{R}$ there exists $\delta > 0$ and $M \in \mathbb{N}$ such that

$$|F_{X_n}(x) - F_N(x)| \leq \varepsilon$$

for $x \in (s - \delta, s + \delta)$ and $n \geq M$. Set $c := \max_{t \in \mathbb{R}} \varphi_N(t)$ and for fixed $s \in \mathbb{R}$ and $\varepsilon > 0$ choose

$$\delta := \frac{\varepsilon}{4c} \, .$$

Furthermore, by the pointwise convergence of F_{X_n} to F_N we find $M \in \mathbb{N}$ such that

$$F_{X_n}(S + \delta) - F_N(s + \delta) \leq \frac{\varepsilon}{2} \text{ and } F_{X_n}(S - \delta) - F_N(s - \delta) \geq -\frac{\varepsilon}{2}$$

for $n \geq M$. Pick any $x \in (s - \delta, s + \delta)$. By the monotonicity of $F_{X_n}(\cdot)$ and $F_N(\cdot)$ we then obtain

$$
\begin{aligned}
F_{X_n}(x) - F_N(x) \quad &\leq \quad F_{X_n}(s + \delta) - F_N(s - \delta) \\
&= \quad F_{X_n}(s + \delta) - F_N(s + \delta) + \int_{s-\delta}^{s+\delta} \varphi(t)dt \leq \tfrac{\varepsilon}{2} + 2\delta\,c = \varepsilon
\end{aligned}
$$

for $n \geq M$. Analogously we verify that

$$
\begin{aligned}
F_{X_n}(x) - F_N(x) \quad &\geq \quad F_{X_n}(s - \delta) - F_N(s + \delta) \\
&= \quad F_{X_n}(s - \delta) - F_N(s - \delta) - \int_{s-\delta}^{s+\delta} \varphi(t)dt \\
&\geq \quad -\tfrac{\varepsilon}{2} - 2\delta\,c = -\varepsilon
\end{aligned}
$$

for $n \geq M$. Hence, combining the last two inequalities we obtain

$$|F_{X_n}(x) - F_N(x)| \leq \varepsilon$$

for $x \in (s - \delta, s + \delta)$ and $n \geq M$, which is the local uniform convergence of F_{X_n} to F_N.

To prove that the convergence is uniform on compact subsets of $\mathbb{R}$, fix any compact subset K of $\mathbb{R}$ and $\varepsilon > 0$. By the local uniform convergence to any $s \in K$ we find an open interval $U_s := (s - \delta(s), s + \delta(s))$ and $M(s) \in \mathbb{N}$ such that

$$|F_{X_n}(x) - F_N(x)| < \varepsilon$$

for $n \geq M(s)$ and $x \in U_s$. The collection $\{U_s | s \in K\}$ forms an open cover of the compact set K. Hence, we find finitely many points $s_1, \ldots, s_m$ in K such that the collection $\{U_{s_j} | j = 1, \ldots, m\}$ is an open cover of K. Setting $\hat{M} := \max\{M(s_1), \ldots, M(s_m)\}$ we find

$$|F_{X_n}(x) - F_N(x)| < \varepsilon$$

for $x \in K$ and $n \geq \hat{M}$. $\square$

13.5   The Theorem of de Moivre–Laplace

In this section we present a special version of the Central Limit Theorem which is known as the Theorem of de Moivre–Laplace. This theorem captures and generalizes the observations made on averages of averages in the above coin tossing experiment in a rigorous mathematical language. The proof of the theorem is quite accessible since only basic tools in calculus need to be applied. A self-contained proof is given in Appendix B.

Rescaling

In the above example we considered sequences of averages

$$S_N = \frac{1}{N} \sum_{i=1}^{N} X_i , \ N = 1, 2, \ldots$$

where $X_1, \ldots, X_N$ were independent and identically distributed random variables. Note that

$$E_P[X_i] = E_P[S_N]$$

for all i. Often it is more convenient to consider the re-scaled sequence

$$S_N^* \stackrel{\text{def}}{=} \frac{\sqrt{N}}{\sigma} (S_N - E_P[S_N]) , \ N = 1, 2, \ldots ,$$

where

$$\sigma^2 = \operatorname{Var}_P[X_i] .$$

This choice transforms the probability distribution of S_N into one centered around 0 and normalizes the variance to 1. Indeed, using the linearity of taking expectations we see that

$$E_P[S_N^*] = 0 .$$

Using the independence of the X_i yields

$$\text{Var}_P[S_N^*] = \frac{N}{\sigma^2} \text{Var}_P[S_N] = \frac{N}{\sigma^2} \frac{1}{N^2} \text{Var}_P\left[\sum_{i=1}^{N} X_i\right] = \frac{N}{\sigma^2} \frac{1}{N^2} N \sigma^2 = 1 .$$

The theorem of de Moivre–Laplace

We now come to the main result of this chapter.

Theorem 13.23 (de Moivre–Laplace) *Consider a sequence $p_n \in (0,1)$ with $p_n \to p \in (0,1)$ and let $(S_n)_{n\in\mathbb{N}}$ denote a sequence of binomial random variables with parameters n and p_n. Then the sequence of re-scaled random variables*

$$S_n^* := \frac{S_n - E_P[S_n]}{\sqrt{\text{Var}(S_n)}} = \frac{S_n - n\,p_n}{\sqrt{n\,p_n\,(1 - p_n)}}$$

converges weakly to the standard normal random variable, i.e.

$$\lim_{n\to\infty} F_{S_n^*}(s) = F_\mathcal{N}(s)$$

for each $s \in \mathbb{R}$.

If the weak convergence of a sequence X_n to the standard normal distribution can be proved, then the following simple corollary is useful for determining the probability distribution of the limit of re-scaled sequences obtained from the weakly convergent variables X_n.

Corollary 13.24 *Let (X_n) be a weakly convergent sequence of random variables with*

$$X_n \overset{w}{\to} \mathcal{N} ,$$

where $\mathcal{N}$ is a standard normal random variable. If (a_n) and (b_n) are sequences in $\mathbb{R}$ with

$$a_n \to a \neq 0 \quad\text{and}\quad b_n \to b$$

then

$$a_n X_n + b_n \overset{w}{\to} \mathcal{N}_{a,b} ,$$

where $\mathcal{N}_{a,b}$ is a normal random variable with variance a^2 and mean b.

Proof Let $s \in \mathbb{R}$. Then, $\frac{s-b_n}{a_n} \to \frac{s-b}{a}$ as n tends to ∞. Hence, by the assumption that $X_n \overset{w}{\to} \mathcal{N}$ and by Lemma 13.22,

$$P(a_n X_n + b_n \leq s) = F_{X_n}\left(\frac{s - b_n}{a_n}\right) \to F_\mathcal{N}\left(\frac{s - b}{a}\right) = P\left(\mathcal{N} \leq \frac{s - b}{a}\right)$$

as n tends to ∞.

Step III: For any random variable $Y : \Omega \to \mathbb{R}$ we obviously have

$$\left|1_{[a,b]}f(Y) - \sum_{k=1}^{r} f(t_i)1_{[t_{i-1},t_i]}\right| \leq \max_{t_{i-1}\leq x,y\leq t_i} |f(x) - f(y)| \leq \frac{\varepsilon}{9}$$

where we have used (13.12).
From this we immediately obtain

$$\left|E_P[1_{[a,b]}f(Y) - \sum_{k=1}^{r} f(t_i)1_{[t_{i-1},t_i]}]\right| \leq \frac{\varepsilon}{9} \tag{13.14}$$

Step IV: We need a last auxiliary estimate.

$$E_P\left[\sum_{k=1}^{r} f(t_i)1_{[t_{i-1}<X_n\leq t_i]} - \sum_{k=1}^{r} f(t_i)1_{[t_{i-1}<X\leq t_i]}\right] \leq M\sum_{k=1}^{r} P(t_{i-1} < X \leq t_i) \leq \frac{\varepsilon}{9}.$$

Step V: We can now finally estimate the middle term of (13.8) by $\frac{\varepsilon}{3}$, completing the proof of the theorem.

$$|E_P[1_{[a<X_n\leq b]}f(X_n) - 1_{[a<X\leq b]}f(X)]|$$

$$\leq |E_P[1_{[a<X_n\leq b]}f(X_n) - \sum_{k=1}^{r} f(t_i)1_{[t_{i-1}<X_n\leq t_i]}]|$$

$$+ |E_P[\sum_{k=1}^{r} f(t_i)1_{[t_{i-1}<X_n\leq t_i]} - \sum_{k=1}^{r} f(t_i)1_{[t_{i-1}<X\leq t_i]}]|$$

$$+ |E_P[\sum_{k=1}^{r} f(t_i)1_{[t_{i-1}<X\leq t_i]} - 1_{[a<X\leq b]}f(X)]|.$$

By Step III the first and third terms on the right-hand side are both smaller than $\frac{\varepsilon}{9}$. By Step IV the second term is also smaller than $\frac{\varepsilon}{9}$. It follows that

$$|E_P[1_{[a<X_n\leq b]}f(X_n) - 1_{[a<X\leq b]}f(X)]| \leq \frac{\varepsilon}{3}. \qquad \square$$

Characterization of weak convergence

Next we prove that the converse of the previous result holds as well, i.e. we show the following characterization of weak convergence: $X_n \overset{w}{\to} X$ is equivalent to the convergence of $E_P[f(X_n)] \to E_P[f(X)]$ for every bounded and continuous function f.

Proposition 13.26 *Assume that $E_P[f(X_n)] \to E_P[f(X)]$ for every bounded and continuous function $f : \mathbb{R} \to \mathbb{R}$. Then $X_n \overset{w}{\to} X$.*

Proof Let x be a point of continuity of F_X. We will show that given $\varepsilon > 0$ we can find $N(\varepsilon) \geq 0$ such that

$$-\varepsilon \leq F_X(x) - F_{X_n}(x) \leq \varepsilon$$

for all $n \geq N(\varepsilon)$. We start by fixing k such that

$$F_X\left(x + \frac{1}{k}\right) - \frac{\varepsilon}{2} \leq F_X(x) \leq F_X\left(x - \frac{1}{k}\right) + \frac{\varepsilon}{2}.$$

This is possible since F_X is continuous in x. We divide the proof into two steps.
Step I: We show that there exists an $N(\varepsilon)$ such that

$$F_X(x) - F_{X_n} \leq \varepsilon$$

for $n \geq N(\varepsilon)$.
Define the function

$$\varphi(t) \stackrel{\text{def}}{=} \begin{cases} 1 & \text{if } t \leq x - \frac{1}{k}, \\ k(x - t) & \text{if } x - \frac{1}{k} \leq t \leq x, \\ 0 & \text{if } t \geq x. \end{cases}$$

The function φ is bounded and continuous. Therefore, $\lim_{n\to\infty} E_P[\varphi(X_n)] = E_P[\varphi(X)]$ and we find $N(\varepsilon)$ such that

$$-\frac{\varepsilon}{2} \leq E_P[\varphi(X_n)] - E_P[\varphi(X)] \leq \frac{\varepsilon}{2}.$$

Note that from the definition of φ we obtain

$$1_{[X \leq x - \frac{1}{k}]} \leq \varphi(X)$$

and

$$\varphi(X_n) \leq 1_{[X_n \leq x]}.$$

It follows that

$$\begin{aligned} F_X\left(x - \frac{1}{k}\right) &= P\left(X \leq x - \frac{1}{k}\right) = E_P[1_{[X \leq x - \frac{1}{k}]}] \\ &\leq E_P[\varphi(X)] \leq E_P[\varphi(X_n)] + \frac{\varepsilon}{2} \\ &= E_P[1_{[X_n \leq x]}] + \frac{\varepsilon}{2} = P([X_n \leq x]) + \frac{\varepsilon}{2} \\ &= F_{X_n} + \frac{\varepsilon}{2}. \end{aligned}$$

From this and the definition of k we obtain

$$F_X(x) \leq F_X\left(x - \frac{1}{k}\right) + \frac{\varepsilon}{2} \leq F_{X_n} + \varepsilon$$

as claimed.

Step II: In a similar way we show that there exists an $N(\varepsilon)$ such that

$$F_X(x) - F_{X_n} \geq -\varepsilon$$

for $n \geq N(\varepsilon)$. Define the function

$$\psi(t) \stackrel{\text{def}}{=} \begin{cases} 1 & \text{if } t \leq x, \\ k(x + \frac{1}{k} - t) & \text{if } x \leq t \leq x + \frac{1}{k}, \\ 0 & \text{if } t \geq x + \frac{1}{k}. \end{cases}$$

The function ψ is bounded and continuous. Therefore, $\lim_{n\to\infty} E_P[\psi(X_n)] = E_P[\psi(X)]$ and we find $N(\varepsilon)$ such that

$$-\frac{\varepsilon}{2} \leq E_P[\psi(X_n)] - E_P[\psi(X)] \leq \frac{\varepsilon}{2} .$$

Note that from the definition of ψ we obtain

$$1_{[X_n \leq x]} \leq \psi(X_n)$$

and

$$\psi(X) \leq 1_{[X_n \leq x + \frac{1}{k}]} .$$

It follows that

$$\begin{aligned} F_{X_n}(x) &= P(X_n \leq x) = E_P[1_{[X_n \leq x]}] \\ &\leq E_P[\psi(X_n)] \leq E_P[\psi(X)] + \frac{\varepsilon}{2} \\ &= E_P[1_{[X \leq x + \frac{1}{k}]}] + \frac{\varepsilon}{2} = P([X_n \leq x + \frac{1}{k}]) + \frac{\varepsilon}{2} \\ &= F_X(x + \frac{1}{k}) + \frac{\varepsilon}{2} . \end{aligned}$$

From this and the definition of k we obtain

$$F_X(x) \geq F_X(x + \frac{1}{k}) - \frac{\varepsilon}{2} \geq F_{X_n} - \varepsilon$$

as claimed.   □

Concluding Remarks and Suggestions for Further Reading

This is one of the simplest versions of a type of theorem proving weak convergence of a sequence of random variables to the normal distribution. More general results of this type can be found for instance in [8]. The theorem of De Moivre–Laplace is crucial for obtaining the Black–Scholes option pricing formula from the Cox–Ross–Rubinstein.

Chapter 14

The Black–Scholes Formula

Unfortunately, the mathematical tools employed in the Black–Scholes and Merton articles are quite advanced and have tended to obscure the underlying economics. However, thanks to a remark by William Sharpe it is possible to derive the same results using only elementary mathematics.

J.C. Cox, S.A. Ross and M. Rubinstein

An introduction to mathematical finance would not be complete without an exposition of its most famous result: the Black–Scholes formula for the price of European call and put options.

14.1 Limiting Behavior of a Cox–Ross–Rubinstein Economy

We consider the Cox–Ross–Rubinstein economy over the time interval $[0, T]$ as parametrized in Section 12.2. For each $N \in \mathbb{N}$ trading activity may take place at the dates

$$0, \ \frac{T}{N}, \ 2\frac{T}{N}, \ \ldots, \ (N-1)\frac{T}{N}, \ T .$$

By choosing $\frac{T}{N}$ as a unit of time trading takes place at the dates $t = 0, 1, \ldots, N$. The underlying sample space is

$$\Omega \stackrel{\text{def}}{=} \{0, 1\}^N .$$

On Ω the sequence $(Z_{N,s})$ of independent random variables is given by

$$Z_{N,s}(\omega) = \omega_s$$

for $\omega = (\omega_1, \ldots, \omega_N)$. Setting,

$$D_{N,t} = \sum_{s=1}^{t} Z_{N,s}$$

we have

$$P(Z_{N,s} = 1) = p_N \qquad \text{and} \qquad P(Z_{N,s} = 0) = 1 - p_N \ ,$$

where p_N is the probability that in one of the subperiods of the model the economic development will be "good".

Given the risk-free rate r_N and the stock yields $y_{g,N}$ and $y_{b,N}$ the money market account at time t is given by

$$B_{N,t} = (1 + r_N)^t,$$

and the price of the stock by

$$S_{N,t}(\omega) = S_0 \left(1 + y_{g,N}\right)^{D_{N,t}(\omega)} \left(1 + y_{b,N}\right)^{t - D_{N,t}(\omega)} . \tag{14.1}$$

Parametrization

As in Section 12.2 we use the expression

$$S_{N,N}(\omega) = S_0 e^{H_N(\omega)}$$

with

$$H_N(\omega) \stackrel{\text{def}}{=} \log \frac{S_{N,N}(\omega)}{S_0(\omega)}.$$

Also in Section 12.2 we showed that, given R such that e^{-RT} is the price of a zero-bond with face value 1 currency unit and maturity T, and given historical observations of

$$\mu \stackrel{\text{def}}{=} \frac{1}{T} E_{P_N} [\log \frac{S_{N,N}}{S_0}] \qquad \text{and} \qquad \sigma^2 \stackrel{\text{def}}{=} \frac{1}{T} Var_{P_N} [\log \frac{S_{N,N}}{S_0}],$$

one can parametrize the Cox–Ross–Rubinstein model for large $N \in \mathbb{N}$ by defining r_N, $y_{g,N}$, $y_{b,N}$ and p_N as

$$r_N \stackrel{\text{def}}{=} e^{\frac{T}{N}R} - 1 \ ,$$

$$y_{g,N} \stackrel{\text{def}}{=} e^{\sqrt{\frac{T}{N}}\sigma} - 1 \ ,$$

$$y_{b,N} \stackrel{\text{def}}{=} e^{-\sqrt{\frac{T}{N}}\sigma} - 1 \ ,$$

$$p_N \stackrel{\text{def}}{=} \frac{1}{2} + \frac{1}{2}\frac{\mu}{\sigma}\sqrt{\frac{T}{N}} \ .$$

We will need the following auxiliary results.

Lemma 14.1 *The following statements hold:*

$$
\begin{aligned}
H_N &= D_{N,N} \log(1 + y_{g,N}) + (N - D_{N,N}) \log(1 + y_{b,N}) , \\
E_{Q_N}[H_N] &= (2q_N - 1)\sigma\sqrt{NT} , \\
Var_{Q_N}[H_N] &= q_N(1 - q_N)4\sigma^2 T .
\end{aligned}
$$

Proof The expression for H_N follows immediately from the definition (14.1) of $S_{N,N}$. To compute the expectation of H_N just note that

$$
\begin{aligned}
E_{Q_N}[H_N] &= E_{Q_N}[D_{N,N}] \log(1 + y_{g,N}) + E_{Q_N}[N - D_{N,N}] \log(1 + y_{b,N}) \\
&= N q_N \log(1 + y_{g,N}) + N(1 - q_N) \log(1 + y_{b,N}) \\
&= N q_N \sqrt{\frac{T}{N}}\sigma - N(1 - q_N)\sqrt{\frac{T}{N}}\sigma \\
&= (2 q_N - 1)\sigma\sqrt{NT} .
\end{aligned}
$$

The variance of H_N is obtained as follows:

$$
\begin{aligned}
Var_{Q_N}[H_N] &= Var_{Q_N}[D_{N,N}][\log(1 + y_{g,N}) - \log(1 + y_{b,N})]^2 \\
&= N q_N(1 - q_N)[\log(1 + y_{g,N}) - \log(1 + y_{b,N})]^2 \\
&= N q_N(1 - q_N)4\sigma^2 \frac{T}{N} \\
&= q_N(1 - q_N)4\sigma^2 T .
\end{aligned}
$$

$\square$

Risk-neutral probability

Since we are interested in looking at prices of contingent claims we need to consider the risk neutral probability. As shown in Theorem 12.7 it is given by

$$
q_N = \frac{r_N - y_{b,N}}{y_{g,N} - y_{b,N}} = \frac{e^{\frac{T}{N}R} - e^{-\sqrt{\frac{T}{N}}\sigma}}{e^{\sqrt{\frac{T}{N}}\sigma} - e^{-\sqrt{\frac{T}{N}}\sigma}} . \tag{14.2}
$$

Lemma 14.2 *The following holds:*

$$
\lim_{N\to\infty} q_N = \frac{1}{2} . \tag{14.3}
$$

Proof Applying de l'Hopital's rule we easily see that for any $a \in \mathbb{R}$,

$$
\lim_{s\to 0} \frac{e^{s^2 a} - e^{-s}}{e^s - e^{-s}} = \frac{1}{2} .
$$

As a consequence of 14.2 the assertion immediately follows by using the substitution $s \stackrel{def}{=} \sqrt{\frac{T}{N}}\sigma$. $\square$

Limiting behavior

We now study the limiting behavior of the Cox–Ross–Rubinstein stock price model under the risk-neutral probability. We start with the following result.

Lemma 14.3 *The following two statements hold:*

$$\lim_{N\to\infty} E_{Q_N}[H_N] = RT - \frac{1}{2}\sigma^2 T$$

and

$$\lim_{N\to\infty} Var_{Q_N}[H_N] = T\sigma^2 .$$

Proof We start by studying the expected value $E_{Q_N}[H_N]$ as N tends to infinity. From Lemma 14.1 we obtain

$$
\begin{aligned}
E_{Q_N}[H_N] &= (2\,q_N - 1)\,\sigma\,\sqrt{NT} \\
&= \frac{2e^{\frac{T}{N}R} - e^{\sqrt{\frac{T}{N}}\sigma} - e^{-\sqrt{\frac{T}{N}}\sigma}}{\sqrt{\frac{T}{N}}\sigma(e^{\sqrt{\frac{T}{N}}\sigma} - e^{-\sqrt{\frac{T}{N}}\sigma})}\,\sigma^2\,T .
\end{aligned}
$$

By applying de l'Hopital's rule twice it is easy to verify that, for any $a \in \mathbb{R}$,

$$\lim_{s\to0} \frac{2e^{s^2 a} - e^s - e^{-s}}{s(e^s - e^{-s})} = a - \frac{1}{2} .$$

It follows that using the substitution $s \stackrel{\text{def}}{=} \sqrt{\frac{T}{N}}\sigma$ we immediately obtain

$$\frac{2e^{\frac{T}{N}R} - e^{\sqrt{\frac{T}{N}}} - e^{-\sqrt{\frac{T}{N}}}}{\sigma\sqrt{\frac{T}{N}}(e^{\sqrt{\frac{T}{N}}} - e^{-\sqrt{\frac{T}{N}}})} \to \frac{R}{\sigma^2} - \frac{1}{2}$$

as N tends to infinity. Therefore,

$$\lim_{N\to\infty} E_{Q_N}[H_N] = RT - \frac{1}{2}\sigma^2 T .$$

Next we study $Var_{Q_N}[H_N]$ as N tends to infinity. From Lemma 14.1 we get

$$Var_{Q_N}[H_N] = q_N(1 - q_N)\,4\sigma^2\,T .$$

We can use (14.3) to see that

$$\lim_{N\to\infty} Var_{Q_N}[H_N] = \sigma^2\,T .$$

□

Applying the central limit theorem

The crucial step in considering the limit $N \to \infty$ is contained in the following proposition. It is an immediate application of the central limit theorem of statistics.

Lemma 14.4 *The random variables*

$$H_N^* \stackrel{def}{=} \frac{H_N - E_{Q_N}[H_N]}{\sqrt{Var_{Q_N}[H_N]}}$$

converge weakly to a standard random variable.

Proof Using Lemma 14.1 it is easy to check that

$$H_N^* = D_{N,N}^* \,,$$

where

$$D_{N,N}^* \stackrel{def}{=} \frac{D_{N,N} - E_{Q_N}[D_{N,N}]}{\sqrt{Var_{Q_N}[D_{N,N}]}} \,.$$

Note that $D_{N,N}$ is Bernoulli distributed with parameters N and q_N, with $q_N \to \frac{1}{2}$ as N tends to infinity. Hence, by the de Moivre–Laplace Theorem 13.23, the sequence $(D_{N,N}^*)$, and thus also (H_N^*), converges weakly to a standard random variable. $\qquad\square$

As a consequence we obtain the following important result.

Proposition 14.5 *The sequence of random variables (H_N) converges (weakly) towards a normally distributed random variable H_T with mean $E[H_T] = RT - \frac{1}{2}\sigma^2 T$ and variance $Var[H_T] = \sigma^2 T$.*

Proof Setting

$$a_N \stackrel{def}{=} \sqrt{Var_{Q_N}[H_N]} \qquad and \qquad b_N \stackrel{def}{=} E_{Q_N}[H_N] \,,$$

we know from Lemma 14.3 that

$$a_N \to \sqrt{T}\sigma \qquad and \qquad b_N \to RT - \frac{1}{2}\sigma^2 T \,.$$

Then we can apply Corollary 13.24 to obtain that

$$H_N = a_N H_N^* + b_N$$

converges weakly to a normal variable H_T with mean $RT - \frac{1}{2}\sigma^2 T$ and variance $\sigma^2 T$. $\qquad\square$

14.2 The Black–Scholes Formula

We are particularly interested in the prices of European call and put options. Recall that these claims are defined by

$$X_{\text{call}} = \max\{ S_T - K , 0\} \quad \text{and} \quad X_{\text{put}} = \max\{ K - S_T , 0\} ,$$

respectively.

In the context of the N-period Cox–Ross–Rubinstein model the price of a call and a put are given by

$$
\begin{aligned}
\pi_0^N[X_{\text{put}}] &= \frac{1}{(1+r_N)^N} E_{Q_N}[\max\{K - S_{N,N} , 0\}] \\
&= e^{-RT} E_{Q_N}[\max\{K - S_0 e^{H_N} , 0\}] .
\end{aligned}
$$

Recall from Proposition 12.12 the put-call parity

$$\pi_0^N[X_{\text{call}}] = \pi_0^N[X_{\text{put}}] + S_0 - e^{-RT}K . \tag{14.4}$$

The put-call parity allows us to deduce the price of a call when the price of the corresponding put is known and vice versa.

Limiting behavior

The objective of the remainder of this section is to study the limit $N \to \infty$ for the Cox–Ross–Rubinstein price of a European put option, i.e. we consider

$$\lim_{N \to \infty} \pi_0^N[X_{\text{put}}] = \lim_{N \to \infty} e^{-RT} E_{Q_N}[\max\{K - S_0 e^{H_N} , 0\}] .$$

The first step is to characterize this limit in terms of an integral over the real line.

Lemma 14.6 *The price of a put option has the limit*

$$\lim_{N \to \infty} \pi_0^N[X_{put}] = e^{-RT} E[\max\{K - S_0 e^{H_T} , 0\}] ,$$

where H_T is the normally distributed random variable introduced in Proposition 14.5.

Proof The function

$$f : \mathbb{R} \to \mathbb{R} , \quad f(x) \stackrel{\text{def}}{=} \max\{K - S_0 e^x , 0\}$$

is continuous and bounded. Hence, we can apply Theorem 13.25 to obtain the result. □

The next result is one of the cornerstones of finance.

Theorem 14.7 (The Black–Scholes Formula) *The price of a European put option in the limit* $N \to \infty$ *is given by*

$$\lim_{N \to \infty} \pi_0^N [X_{put}] = K e^{-RT} F_{\mathcal{N}}(-d_2) - S_0 F_{\mathcal{N}}(-d_1) ,$$

where

$$d_1 \stackrel{def}{=} \frac{\ln(\frac{S_0}{K} e^{RT}) + \frac{\sigma^2}{2} T}{\sigma \sqrt{T}} \quad , \quad d_2 \stackrel{def}{=} d_1 - \sigma \sqrt{T}$$

and

$$F_{\mathcal{N}}(x) \stackrel{def}{=} \frac{1}{\sqrt{2\pi}} \int_{-\infty}^{x} e^{-\frac{y^2}{2}} \, dy$$

is the probability distribution of the standard normal distribution.
The price of a European call option in the limit $N \to \infty$ *is given by*

$$\lim_{N \to \infty} \pi_0^N [X_{call}] = S_0 F_{\mathcal{N}}(d_1) - K e^{-RT} F_{\mathcal{N}}(d_2) .$$

Proof By Lemma 14.6 the price of a European put option in the limit $N \to \infty$ is given by

$$P := e^{-RT} E \left[\max \left\{ K - S_0 e^{H_T} , 0 \right\} \right] .$$

By Theorem 13.17 we have

$$E \left[\max \left\{ K - S_0 e^{H_T} , 0 \right\} \right] = \frac{1}{\sigma \sqrt{2\pi T}} \int_{-\infty}^{\infty} f(y) e^{-\frac{1}{2} \left(\frac{y - RT + \frac{1}{2}\sigma^2 T}{\sigma \sqrt{T}} \right)^2} \, dy , \quad (14.5)$$

where

$$f(y) = \max \left\{ K - S_0 e^y , 0 \right\} .$$

Making the substitution of variables

$$x \stackrel{def}{=} \frac{y - RT + \frac{\sigma^2}{2} T}{\sigma \sqrt{T}} ,$$

we find, since $dx = \frac{dy}{\sigma \sqrt{T}}$, that

$$P = \frac{e^{-RT}}{\sqrt{2\pi}} \int_{-\infty}^{\infty} f(\sigma \sqrt{T} x + RT - \frac{\sigma^2}{2} T) e^{-\frac{x^2}{2}} \, dx .$$

Inserting $f(y)$ yields

$$P = \frac{e^{-RT}}{\sqrt{2\pi}} \int\limits_{-\infty}^{\infty} \max\left\{ K - S_0 e^{\left(\sigma\sqrt{T}\,x + RT - \frac{\sigma^2}{2}T\right)}, 0 \right\} e^{-\frac{x^2}{2}}\, dx \ .$$

Since $f(y)$ is non-zero only when

$$K - S_0 e^{\left(\sigma\sqrt{T}\,x + RT - \frac{\sigma^2}{2}T\right)} > 0 \ ,$$

or, equivalently, when

$$x < -\frac{\ln\left(\frac{S_0}{K} e^{RT}\right) - \frac{\sigma^2}{2}T}{\sigma\sqrt{T}} = -d_2 \ ,$$

we conclude that

$$
\begin{aligned}
P &= \frac{e^{-RT}}{\sqrt{2\pi}} \int\limits_{-\infty}^{-d_2} \left[K - S_0 e^{\left(\sigma\sqrt{T}\,x + RT - \frac{\sigma^2}{2}T\right)} \right] e^{-\frac{x^2}{2}}\, dx \\
&= \frac{1}{\sqrt{2\pi}} \int\limits_{-\infty}^{-d_2} \left[K e^{-RT} - S_0 e^{\left(\sigma\sqrt{T}\,x - \frac{\sigma^2}{2}T\right)} \right] e^{-\frac{x^2}{2}}\, dx \ .
\end{aligned}
$$

Since

$$\sigma\sqrt{T}\,x - \frac{\sigma^2}{2}T - \frac{x^2}{2} = -\frac{1}{2}(x - \sigma\sqrt{T})^2 \ ,$$

we see that

$$
\begin{aligned}
P &= \frac{1}{\sqrt{2\pi}} K e^{-RT} \int\limits_{-\infty}^{-d_2} e^{-\frac{x^2}{2}}\, dx - \frac{1}{\sqrt{2\pi}} S_0 \int\limits_{-\infty}^{-d_2} e^{\left(\sigma\sqrt{T}\,x - \frac{\sigma^2}{2}T\right)} e^{-\frac{x^2}{2}}\, dx \\
&= \frac{1}{\sqrt{2\pi}} K e^{-RT} \int\limits_{-\infty}^{-d_2} e^{-\frac{x^2}{2}}\, dx - \frac{1}{\sqrt{2\pi}} S_0 \int\limits_{-\infty}^{-d_2} e^{-\frac{1}{2}(x - \sigma\sqrt{T})^2}\, dx \ .
\end{aligned}
$$

Using the change of variables $y = x - \sigma\sqrt{T}$ in the second integral and noting that $d_1 = d_2 + \sigma\sqrt{T}$ results in

$$
\begin{aligned}
P &= \frac{1}{\sqrt{2\pi}} K e^{-RT} \int\limits_{-\infty}^{-d_2} e^{-\frac{x^2}{2}}\, dx - \frac{1}{\sqrt{2\pi}} S_0 \int\limits_{-\infty}^{-d_2 - \sigma\sqrt{T}} e^{-\frac{x^2}{2}}\, dx \\
&= K e^{-RT} F_{\mathcal{N}}(-d_2) - S_0 F_{\mathcal{N}}(-d_1) \ .
\end{aligned}
$$

The formula for the European call can be obtained by applying the put-call parity (14.4). In fact, for each $N \in \mathbb{N}$ the put-call parity implies

$$\pi_0^N[X_{\mathrm{call}}] = \pi_0^N[X_{\mathrm{put}}] + S_0 - \frac{K}{(1+r_N)^N} = \pi_0^N[X_{\mathrm{put}}] + S_0 - \frac{K}{e^{RT}} .$$

Passing to the limit on both sides of this equation and inserting the above formula for the price of the put option leads to

$$\begin{aligned}
\lim_{N\to\infty} \pi_0^N[X_{\mathrm{call}}] &= S_0(1 - F_{\mathcal{N}}(-d_1)) + Ke^{-RT}(F_{\mathcal{N}}(-d_2) - 1) \\
&= S_0 \, F_{\mathcal{N}}(d_1) - K\, e^{-RT} \, F_{\mathcal{N}}(d_2) .
\end{aligned}$$

$\square$

Concluding Remarks and Suggestions for Further Reading

The celebrated Black–Scholes formula was proved in [9] and [43]. The mathematical tools used in these original articles are, however, very sophisticated. The simple derivation from the Cox–Ross–Rubinstein model represented a major break through and was given in [17]. The Black–Scholes formula is still one of the most widely used formulas in practice.

An intuitive (but not so rigorous) treatment can be found in [29]. More mathematical treatments can be found in [5], [20], [39] or [45].

Chapter 15

Optimal Stopping

If the gambler can foresee the future, he would not need probability theory! In this sense an optional [stopping] time has also been described as being "independent of the future"; it must have been decided upon as an "option" without the advantage of clairvoyance.

K.L. Chung

15.1 Stopping Times Introduced

In this chapter we formalize the notion of a rule for deciding when to stop playing a game. This type of problem will arise in a natural way in connection with the optimal exercise of American options.

Assume we are given a sample space $\Omega = \{\omega_1, \ldots, \omega_n\}$, a probability measure P on Ω and an information structure $\mathcal{P} = (\mathcal{P}_t)$. Recall that information can be alternatively described by the family $(\mathcal{A}_t)$, where $\mathcal{A}_t \overset{\text{def}}{=} \mathcal{A}(\mathcal{P}_t)$ is the algebra generated by $\mathcal{P}_t$.

Consider now an adapted process (Z_t). It will be useful to keep the following situation in mind: a game is played in which at any time t the player has the option either to quit the game, receiving the reward Z_t, or continue playing.

A stopping rule should tell us at each time t whether to quit the game or not, conditional on the information available to us at that time. Formally, a *stopping rule* or a *stopping time* with respect to $\mathcal{P}$ is a random variable

$$\tau : \Omega \to \{0, 1, \ldots, T\}$$

such that

$$[\tau = t] = [\omega \in \Omega \, ; \, \tau(\omega) = t] \in \mathcal{A}_t \, . \tag{15.1}$$

Condition (15.1) ensures that when we receive the information $\mathcal{P}_t$ we will be able to tell whether $[\tau = t]$ has occurred or not, i.e. we will know whether to stop or not.

A stopping time is a particular type of *random time*, the latter being just a random variable $\nu : \Omega \to \{0, 1, \ldots, T\}$ not necessarily satisfying the condition (15.1).

Examples 15.1 (a) Let $B \subset \mathbb{R}$ be given. Then,

$$\tau_B^{in}(\omega) \stackrel{\text{def}}{=} \begin{cases} \min\{t;\ Z_t(\omega) \in B\} & \text{if } \{t;\ Z_t(\omega) \in B\} \neq \emptyset, \\ T & \text{otherwise}, \end{cases}$$

is called the *first entry time* of (Z_t) into B. The function τ_B^{in} is a stopping time. Indeed, if $t < T$ we have

$$[\tau_B^{in} = t] = ([Z_0 \in B^c] \cap \ldots \cap [Z_{t-1} \in B^c]) \cap [Z_t \in B].$$

Since for all $s \leq t$ both $[Z_s \in B]$ and $[Z_s \in B^c]$ belong to $\mathcal{A}_s \subset \mathcal{A}_t$, we obtain that $[\tau_B^{in} = t]$ belongs to $\mathcal{A}_t$. For $t = T$ we have

$$[\tau_B^{in} = T] = [Z_0 \in B^c] \cap \ldots \cap [Z_{T-1} \in B^c] \in \mathcal{A}_{T-1}.$$

This proves that τ_B^{in} is a stopping time. The other side of the coin is shown by the *first exit time* of (Z_t) from B defined by

$$\tau_B^{out}(\omega) \stackrel{\text{def}}{=} \begin{cases} \min\{t;\ Z_t(\omega) \notin B\} & \text{if } \{t;\ Z_t(\omega) \notin B\} \neq \emptyset, \\ T & \text{otherwise}. \end{cases}$$

Since the first exit time from B is equal to the first entry time into B^c we see that τ_B^{out} is a stopping time.

(b) Let $(S_t)_{t=0,1,\ldots,T}$ describe the price process of a stock. Then,

$$\tau(\omega) \stackrel{\text{def}}{=} \begin{cases} \min\{t;\ S_t(\omega) \geq 2S_0\} & \text{if } \{t;\ S_t(\omega) \geq 2S_0\} \neq \emptyset, \\ T & \text{otherwise}, \end{cases}$$

describes the first time that the stock price doubles its initial price. This is the first entry time of (S_t) into $A \stackrel{\text{def}}{=} [2S_0, \infty)$ and thus τ is a stopping time.

(c) Another example of a stopping time is the first time at which the stock suffers a negative return, i.e.

$$\tau(\omega) \stackrel{\text{def}}{=} \begin{cases} \min\{t;\ \frac{S_t(\omega)}{S_{t-1}(\omega)} < 1\} & \text{if } \{t;\ \frac{S_t(\omega)}{S_{t-1}(\omega)} < 1\} \neq \emptyset, \\ T & \text{otherwise}. \end{cases}$$

This is the first entry time of $\left(\frac{S_t}{S_{t-1}}\right)$ into $(-\infty, 1)$ and, hence, a stopping time.

(d) Consider now the time just before the first negative return is realized, i.e.

$$\tau(\omega) \stackrel{\text{def}}{=} \begin{cases} \min\{t;\ \frac{S_{t+1}(\omega)}{S_t(\omega)} < 1\} & \text{if } \{t;\ \frac{S_{t+1}(\omega)}{S_t(\omega)} < 1\} \neq \emptyset. \\ T & \text{otherwise}. \end{cases}$$

To know whether we should stop or not at time t we would need to know whether $\frac{S_{t+1}(\omega)}{S_t(\omega)} < 1$. Thus, at time t we need information on the value of the stock at time $t+1$, a piece of information which is only available at time $t+1$. It follows that $[\frac{S_{t+1}(\omega)}{S_t(\omega)} < 1] \notin \mathcal{A}_t$. Therefore, τ is a random time but not a stopping time.

The above example illustrates quite clearly that stopping times are random times that do not require knowledge about the future.

Simple properties

For any $s \le t$ define

$$\mathcal{S}_{s,t} \stackrel{\text{def}}{=} \{\tau \; ; \; \tau \text{ is a stopping time and } s \le \tau \le t\} \, .$$

Hence, $\mathcal{S}_{s,t}$ is the collection of all stopping times by which a process is stopped no earlier than s and no later than t. Obviously $\mathcal{S}_{0,T}$ is the set of all stopping times.

Remark 15.2 *It is worth noting that since both Ω and $\{s, s+1, \ldots, t\}$ are finite sets, the collection $\mathcal{S}_{s,t}$ is also finite.*

We collect a few elementary properties of stopping times in the following lemma. The first property is that at any time t we know whether the process has already been stopped at an earlier time or not. The second property is that the property of being a stopping time is preserved when building minima, maxima, and sums of stopping times.

Lemma 15.3   *a) A random time $\tau : \Omega \to \{0, 1, \ldots, T\}$ is a stopping time if and only if $[\tau \le t] \in \mathcal{A}_t$ for all $t \in \{0, 1, \ldots, T\}$.*

b) If τ_1 and τ_2 are stopping times in $\mathcal{S}_{s,t}$ ($s \le t$) so are $\tau_m := \min\{\tau_1, \tau_2\}$, $\tau_M := \max\{\tau_1, \tau_2\}$, and $\tau_1 + \tau_2$.

Proof a) Assume τ is a stopping time, i.e. $[\tau = t] \in \mathcal{A}_t$ for all $t \in \{0, 1, \ldots, T\}$. Observe that

$$[\tau \le t] = [\tau = 0] \cup [\tau = 1] \cup \ldots \cup [\tau = t] \, .$$

Since $[\tau = s] \in \mathcal{A}_s \subset \mathcal{A}_t$ for all $s \le t$, we have $[\tau \le t] \in \mathcal{A}_t$. If, on the other hand, $[\tau \le t] \in \mathcal{A}_t$ holds for all t, then we conclude

$$[\tau = t] = [\tau \le t] \cap [\tau \le t-1]^c \in \mathcal{A}_t \, .$$

b) From

$$[\tau_m \le t] = [\tau_1 \le t] \cup [\tau_2 \le t] \in \mathcal{A}_t$$

and

$$[\tau_M \le t] = [\tau_1 \le t] \cap [\tau_2 \le t] \in \mathcal{A}_t$$

we see that τ_m and τ_M are stopping times.

From

$$[\tau_1 + \tau_2 = t] = \bigcup_{s=0}^{t} ([\tau_1 = s] \cap [\tau_2 = t - s])$$

and since both

$$[\tau_1 = s] \in \mathcal{A}_s \subset \mathcal{A}_t$$

and

$$[\tau_2 = t - s] \in \mathcal{A}_{t-s} \subset \mathcal{A}_t$$

hold for $s = 0, \ldots, t$, we find that $[\tau_1 + \tau_2 = t] \in \mathcal{A}_t$. $\square$

15.2   Sampling a Process by a Stopping Time

We now formalize what it means to stop an adapted process (Z_t) by a given stopping time τ. Define

$$Z_t^\tau(\omega) := \begin{cases} Z_t(\omega) & \text{if} \quad t \leq \tau(\omega), \\ Z_{\tau(\omega)}(\omega) & \text{if} \quad t \geq \tau(\omega). \end{cases}$$

The process (Z_t^τ) is called the *stopped* or *sampled* process. We may write the stopped process as

$$Z_t^\tau = 1_{[\tau \geq t]} Z_t + \sum_{s=0}^{t-1} 1_{[\tau=s]} Z_s \tag{15.2}$$

as is easily checked.

Lemma 15.4 *The stopped process (Z_t^τ) is adapted. If (Z_t) is predictable so is (Z_t^τ).*

Proof Let $t \in \{0, \ldots, T\}$ then

$$[\tau \geq t] = [\tau \leq t - 1]^c \in \mathcal{A}_{t-1}$$

and

$$[\tau = s] \in \mathcal{A}_s \subset \mathcal{A}_{t-1}$$

for all $s = 0, \ldots, t - 1$. We immediately see that $1_{[\tau \geq t]}$ and $1_{[\tau=s]}$ are all $\mathcal{P}_{t-1}$-measurable. If (Z_t) is predictable it then follows that $1_{[\tau \geq t]} Z_t$ and $1_{[\tau=s]} Z_s$ are all $\mathcal{P}_{t-1}$-measurable. The predictability of (Z_t^τ) follows from (15.2). In case that (Z_t) is only adapted we can only infer that $1_{[\tau \geq t]} Z_t$ and $1_{[\tau=s]} Z_s$ are $\mathcal{P}_t$-measurable. This yields the adaptedness of (Z_t^τ). $\square$

The following result states that sampling a process is order preserving. Its proof is obvious.

Lemma 15.5 *If the process (U_t) dominates the process (Z_t), i.e. if $U_t \geq Z_t$ for all t, then stopped process (U_t^τ) dominates the stopped process (Z_t^τ).*

Doob's optional sampling theorem

The following result shows that if we stop a martingale by a stopping time the stopped process is still a martingale.

Theorem 15.6 *Let (Z_t) be a martingale and τ a stopping time. Then (Z_t^τ) is also a martingale. The same is true if we replace "martingale" by "sub-" or "supermartingale".*

Proof We show the validity of the theorem for submartingales. The other assertions are proved analogously. We need to show that for any $t \in \{0, \ldots, T-1\}$

$$E_P[Z_{t+1}^\tau] \geq Z_t^\tau \ .$$

By (15.2) we have

$$E_P[Z_{t+1}^\tau | \mathcal{P}_t] = E_P[1_{[\tau \geq t+1]} Z_{t+1} | \mathcal{P}_t] + \sum_{s=0}^t E_P[1_{[\tau=s]} Z_s | \mathcal{P}_t] \ .$$

Since $1_{[\tau=s]} Z_s$ is $\mathcal{P}_t$-measurable for all $s = 0, \ldots, t$ we have

$$E_P[1_{[\tau=s]} Z_s | \mathcal{P}_t] = 1_{[\tau=s]} Z_s \ .$$

Moreover, since $1_{[\tau \geq t+1]}$ is $\mathcal{P}_t$-measurable,

$$E_P[1_{[\tau \geq t+1]} Z_{t+1} | \mathcal{P}_t] = 1_{[\tau \geq t+1]} E_P[Z_{t+1} | \mathcal{P}_t] \ .$$

Hence,

$$
\begin{aligned}
E_P[Z_{t+1}^\tau | \mathcal{P}_t] &= 1_{[\tau \geq t+1]} E_P[Z_{t+1} | \mathcal{P}_t] + \sum_{s=0}^t 1_{[\tau=s]} Z_s \\
&\geq 1_{[\tau \geq t+1]} Z_t + \sum_{s=0}^t 1_{[\tau=s]} Z_s \\
&= 1_{[\tau \geq t+1]} Z_t + 1_{[\tau=t]} Z_t + \sum_{s=0}^{t-1} 1_{[\tau=s]} Z_s \\
&= 1_{[\tau \geq t]} Z_t + \sum_{s=0}^{t-1} 1_{[\tau=s]} Z_s \\
&= Z_t^\tau ,
\end{aligned}
$$

where we have used the submartingale property for the inequality. $\square$

The above result tells us that no matter which stopping strategy we use, a fair game will remain fair. Similarly, an unfavorable game cannot turn favorable and a favorable game cannot turn unfavorable by the choice of a stopping rule. It is however easily seen that an (un)favorable game can turn fair by an appropriate stopping strategy. For example consider a coin tossing game where first a fair coin is tossed and subsequently the fair coin is replaced by an unfavorable one. Obviously this game is unfavorable if played to the end. However, a stopping strategy which turns this game into a fair one consists in quitting the game before the unfair coin starts to be used.

Terminal or final value of a stopped process

The random variable

$$Z_\tau(\omega) \stackrel{\text{def}}{=} Z_{\tau(\omega)}(\omega)$$

is called the *terminal* or *final* value of the stopped process because, as is easily seen, $Z_{\tau(\omega)}(\omega)$ indeed coincides with the terminal value $Z_T^\tau(\omega)$ of the process (Z_t^τ).
Note that if $\tau \in \mathcal{S}_{s,t}$, then the process will be stopped the very latest at time t so that in this case $Z_\tau = Z_t^\tau$ holds. Therefore, Z_τ is the variable which the process will eventually "converge" to.
The following result tells us that if (Z_t) is a martingale, i.e. if the game is fair, the expected terminal gain when following any stopping strategy τ is equal to the initial wealth. If on the other hand, (Z_t) is a submartingale, i.e. if the game is favorable, the expected terminal gain when following any stopping strategy τ lies above the initial wealth. Finally, if (Z_t) is a supermartingale, i.e. if the game is unfavorable, the expected terminal gain when following any stopping strategy τ lies below the initial wealth.

Proposition 15.7 *Let (Z_t) be a martingale and τ a stopping time. Then,*

$$E_P[Z_\tau] = Z_0 .$$

If (Z_t) is a submartingale, then

$$E_P[Z_\tau] \geq Z_0 .$$

If (Z_t) is a supermartingale, then

$$E_P[Z_\tau] \leq Z_0 .$$

Proof If (Z_t) is a supermartingale, then Z_t^τ is also a supermartingale by the optional sampling theorem. Therefore, we have

$$E_P[Z_\tau] = E_P[Z_T^\tau] \leq Z_0^\tau = Z_0 .$$

The assertion for submartingales follows similarly. The assertion for martingales follows from the fact that a martingale is both a supermartingale and a submartingale. $\square$

15.3   Optimal Stopping

Now that we have seen how to formalize the concept of a stopping rule, it is natural to ask whether or not it is possible to find an optimal one. The precise formulation of optimality follows.

Definition of optimality

A stopping time τ^* is called *optimal* if

$$E_P[Z_{\tau^*}] = \max_{\tau \in \mathcal{S}_{0,T}} E_P[Z_\tau] . \tag{15.3}$$

Remark 15.8 *Note that since by Remark 15.2 the set $\mathcal{S}_{0,T}$ is finite, the maximum always exists. We shall often use this fact without further reference.*

Recall our interpretation of Z_t as the achieved gain, should the player decide to stop at time t. An optimal stopping rule thus corresponds to a strategy which maximizes expected gains.

In this section we characterize optimal stopping times. We start by introducing in an intuitive way a stochastic process which is sometimes called Snell's envelope.

15.3.1 Snell's Envelope

Assume the player has not quit the game up to time t. He still has the choice of all stopping strategies in $\mathcal{S}_{t,T}$. It is clear that, analogously to the optimality criterion (15.3), he should choose a stopping rule τ^* in $\mathcal{S}_{t,T}$ which maximizes expected gains conditional on the information available at that time, i.e. such that

$$E_P[Z_{\tau^*}|\mathcal{P}_t] = \max_{\tau \in \mathcal{S}_{t,T}} E_P[Z_\tau|\mathcal{P}_t].$$

As in (15.3), this maximum exists since $\mathcal{S}_{t,T}$ is a finite set.

Note that stopping at time t means choosing the stopping time $\tau_t \equiv t$ which belongs to $\mathcal{S}_{t,T}$. Therefore, the above implies that stopping at time t only makes sense if

$$Z_t = \max_{\tau \in \mathcal{S}_{t,T}} E_P[Z_\tau|\mathcal{P}_t] , \tag{15.4}$$

where we have used that

$$Z_t = E_P[Z_t|\mathcal{P}_t] = E_P[Z_{\tau_t}|\mathcal{P}_t]$$

holds because Z_t is $\mathcal{P}_t$-measurable.

We now proceed to investigate the right-hand side of equation (15.4). To that effect we set

$$U_T \stackrel{\text{def}}{=} Z_T$$

and

$$U_t \stackrel{\text{def}}{=} \max_{\tau \in \mathcal{S}_{t,T}} E_P[Z_\tau|\mathcal{P}_t].$$

The stochastic process $(U_t)_t$ is called the *Snell envelope* of (Z_t). The Snell envelope thus indicates at each time the gain the player could expect if he were to choose an optimal strategy from that point onwards.

Lemma 15.9 *The Snell envelope of* (Z_t) *is a* $(\mathcal{P}_t)$-*adapted process.*

Proof Note that U_t is defined as the maximum over a finite number of $\mathcal{P}_t$-measurable random variables. It follows from Corollary 7.13 that U_t is measurable with respect to $\mathcal{P}_t$, proving its adaptedness.   □

We next characterize Snell's envelope by a recursion formula.

A characterization of Snell's envelope

Before stating the recursion formula we need the following result. Its interpretation is that in order to determine the optimal expected value of the gains that can be reached by choosing an optimal stopping rule, we can just look at the two alternatives: either stopping immediately or else using a stopping rule which stops at $t+1$ the very earliest.

Lemma 15.10 *For all* t *in* $\{0, \ldots, T-1\}$ *we have*

$$U_t = \max\{Z_t, \max_{\tau \in \mathcal{S}_{t+1,T}} E_P[Z_\tau|\mathcal{P}_t]\}.$$

Proof
(i) Since $\tau_t \equiv t$ belongs to $\mathcal{S}_{t,T}$ we have

$$Z_t = Z_{\tau_t} = E_P[Z_{\tau_t}|\mathcal{P}_t] \leq \max_{\tau \in \mathcal{S}_{t,T}} E_P[Z_\tau|\mathcal{P}_t] = U_t .$$

On the other hand, since $\mathcal{S}_{t+1,T} \subset \mathcal{S}_{t,T}$, we have

$$\max_{\tau \in \mathcal{S}_{t+1,T}} E_P[Z_\tau|\mathcal{P}_t] \leq \max_{\tau \in \mathcal{S}_{t,T}} E_P[Z_\tau|\mathcal{P}_t] = U_t .$$

It follows that

$$\max\{Z_t, \max_{\tau \in \mathcal{S}_{t,T}} E_P[Z_\tau|\mathcal{P}_t]\} \leq U_t .$$

(ii) Take any $\tau \in \mathcal{S}_{t,T}$ and set

$$\nu \stackrel{\text{def}}{=} \max\{\tau, t+1\} \in \mathcal{S}_{t+1,T} .$$

Thus, ν corresponds to stopping according to τ but not earlier than $t+1$. Observe that on the set $[\tau \geq t+1]$ we have that $\nu = \tau$ and consequently $Z_\nu = Z_\tau$ on this set. Moreover, obviously $Z_\tau = Z_t$ on the set $[\tau = t]$. From this we infer that

$$Z_\tau = 1_{[\tau=t]} Z_t + 1_{[\tau \geq t+1]} Z_\tau = 1_{[\tau=t]} Z_t + 1_{[\tau \geq t+1]} Z_\nu .$$

Recall that $\tau \in \mathcal{S}_{t,T}$ so that $[\tau = t]$ and $[\tau \geq t+1] = [\tau = t]^c$ both belong to $\mathcal{P}_t$. We thus conclude

$$
\begin{aligned}
E_P[Z_\tau|\mathcal{P}_t] &= E_P[1_{[\tau=t]}Z_t|\mathcal{P}_t] + E_P[1_{[\tau\geq t+1]}Z_\nu|\mathcal{P}_t] \\
&= 1_{[\tau=t]}Z_t + 1_{[\tau\geq t+1]}E_P[Z_\nu|\mathcal{P}_t] \\
&\leq 1_{[\tau=t]}Z_t + 1_{[\tau\geq t+1]}\max_{\mu\in\mathcal{S}_{t+1,T}}E_P[Z_\mu|\mathcal{P}_t] \\
&\leq 1_{[\tau=t]}\max\{Z_t, \max_{\mu\in\mathcal{S}_{t+1,T}}E_P[Z_\mu|\mathcal{P}_t]\} \\
&\quad + 1_{[\tau\geq t+1]}\max\{Z_t, \max_{\mu\in\mathcal{S}_{t+1,T}}E_P[Z_\mu|\mathcal{P}_t]\} \\
&= \max\{Z_t, \max_{\mu\in\mathcal{S}_{t+1,T}}E_P[Z_\mu|\mathcal{P}_t]\}.
\end{aligned}
$$

Since $\tau \in \mathcal{S}_{t,T}$ was arbitrary and the last expression is independent of τ, this proves that

$$
U_t = \max_{\tau\in\mathcal{S}_{t,T}} E_P[Z_\tau|\mathcal{P}_t] \leq \max\{Z_t, \max_{\mu\in\mathcal{S}_{t+1,T}}E_P[Z_\mu|\mathcal{P}_t]\}.
$$

The assertion now follows from (i) and (ii). □

As a corollary we obtain the recursive characterization of Snell's envelope. This characterization is in fact frequently used as the definition of the Snell envelope. We preferred to state it as a consequence from the more intuitive definition given above.

Proposition 15.11 *For all t in $\{0,\ldots,T\}$ the following recursion formula holds:*

$$
U_t = \max\{Z_t, E_P[U_{t+1}|\mathcal{P}_t]\}.
$$

Proof We show that

$$
E_P[U_{t+1}|\mathcal{P}_t] = \max_{\tau\in\mathcal{S}_{t+1,T}} E_P[Z_\tau|\mathcal{P}_t] \tag{15.5}
$$

which by Lemma 15.10 then yields the assertion.
(i) By the tower property of conditional expectation we have

$$
\begin{aligned}
\max_{\tau\in\mathcal{S}_{t+1,T}} E_P[Z_\tau|\mathcal{P}_t] &= \max_{\tau\in\mathcal{S}_{t+1,T}} E_P[E_P[Z_\tau|\mathcal{P}_{t+1}]\,|\,\mathcal{P}_t] \\
&\leq E_P[\max_{\tau\in\mathcal{S}_{t+1,T}} E_P[Z_\tau|\mathcal{P}_{t+1}]\,|\,\mathcal{P}_t] \\
&= E_P[U_{t+1}|\mathcal{P}_t],
\end{aligned}
$$

where for the inequality we have used Lemma 10.7.
(ii) Let $\tau^* \in \mathcal{S}_{t+1,T}$ be such that

$$
U_{t+1} = \max_{\tau\in\mathcal{S}_{t+1,T}} E_P[Z_\tau|\mathcal{P}_{t+1}] = E_P[Z_{\tau^*}|\mathcal{P}_{t+1}].
$$

Then,

$$
\begin{aligned}
E_P[U_{t+1}|\mathcal{P}_t] &= E_P\big[\,E_P[Z_{\tau^*}|\mathcal{P}_{t+1}]\,|\,\mathcal{P}_t\,\big] \\
&= E_P[Z_{\tau^*}|\mathcal{P}_t] \\
&\leq \max_{\tau \in \mathcal{S}_{t+1,T}} E_P[Z_\tau|\mathcal{P}_t],
\end{aligned}
$$

which concludes the proof of the result. $\qquad\square$

Smallest supermartingale dominating (Z_t)

The following easy consequence of the recursion formula will prove useful later on.

Corollary 15.12 *The Snell envelope (U_t) of (Z_t) is a supermartingale. It is the smallest supermartingale which dominates (Z_t), i.e. such that*

$$
U_t \geq Z_t
$$

for all t in $\{0,\ldots,T\}$.

Proof We already know that (U_t) is $\mathcal{P}_t$-adapted. To see that it is a supermartingale just note that by the previous proposition

$$
U_t = \max\{Z_t, E_P[U_{t+1}|\mathcal{P}_t]\} \geq E_P[U_{t+1}|\mathcal{P}_t].
$$

Assume now that (V_t) is any supermartingale dominating (Z_t), i.e.

$$
V_t \geq Z_t
$$

for $t = 0,\ldots,T$. The supermartingale property implies that

$$
V_t \geq E_P[V_{t+1}|\mathcal{P}_t] .
$$

It follows that

$$
V_t \geq \max\{Z_t, E_P[V_{t+1}|\mathcal{P}_t]\} .
$$

Note that since V_T dominates Z_T and $Z_T = U_T$ holds we have $V_T \geq U_T$. We apply a backward induction argument assuming that $V_{t+1} \geq U_{t+1}$. We have thus to show that $V_t \geq U_t$. This follows from

$$
\begin{aligned}
V_t &\geq \max\{Z_t, E_P[V_{t+1}|\mathcal{P}_t]\} \\
&\geq \max\{Z_t, E_P[U_{t+1}|\mathcal{P}_t]\} \\
&= U_t .
\end{aligned}
$$

$\qquad\square$

15.3.2   A Characterization of Optimal Stopping Times

By the Optional Sampling Theorem a supermartingale when stopped can at best
have the property of being a martingale. It thus seems natural that an optimally
stopped supermartingale should have this property. On the other hand it also
seems reasonable that if a stopping time is optimal, the gains obtained when
following the implied strategy will be larger than or equal to the gains expected
from continuing the game. In fact, optimal stopping times are characterized by
these two properties.

Theorem 15.13 *A stopping time* $\tau \in \mathcal{S}_{0,T}$ *is optimal if and only if the stopped
Snell envelope* (U_t^τ) *is a martingale and*

$$U_\tau = Z_\tau \ .$$

If τ *is an optimal stopping time, then we have*

$$U_0 = E_P[Z_\tau | \mathcal{P}_0] \ .$$

Proof (i) To show sufficiency of the conditions let τ be a stopping time such that
$U_\tau = Z_\tau$ holds and (U_t^τ) is a martingale. This assumption implies

$$U_0 = U_0^\tau = E_P[U_T^\tau | \mathcal{P}_0] = E_P[U_\tau | \mathcal{P}_0] = E_P[Z_\tau | \mathcal{P}_0] \ .$$

From Corollary 15.12 we know that (U_t) is the smallest supermartingale dominating (Z_t). Therefore, by Doob's optional sampling theorem (Theorem 15.6)
and Lemma 15.5, for any $\nu \in \mathcal{S}_{0,T}$ we have that (U_t^ν) is a supermartingale with
$U_t^\nu \geq Z_t^\nu$. We therefore have that

$$E_P[Z_\nu | \mathcal{P}_0] = E_P[Z_T^\nu | \mathcal{P}_0] \leq E_P[U_T^\nu | \mathcal{P}_0] \leq U_0^\nu = U_0$$

holds. It follows that for all $\nu \in \mathcal{S}_{0,T}$ we have

$$E_P[Z_\nu | \mathcal{P}_0] \leq U_0 = E_P[Z_\tau | \mathcal{P}_0] \ ,$$

proving the optimality of τ.
(ii) It remains to show the necessity of the conditions. To that effect let τ be
optimal. We first show that $U_\tau = Z_\tau$. Note first that since (U_t) dominates (Z_t) we
obviously have $Z_\tau \leq U_\tau$. Hence, by the definition of U_0 and by the optimality of
τ we obtain

$$U_0 = E_P[Z_\tau | \mathcal{P}_0] \leq E_P[U_\tau | \mathcal{P}_0] \ .$$

Furthermore, since (U_t^τ) is a supermartingale we obtain

$$U_0 = U_0^\tau \geq E_P[U_T^\tau | \mathcal{P}_0] = E_P[U_\tau | \mathcal{P}_0] \ .$$

These two inequalities yield

$$E_P[U_\tau | \mathcal{P}_0] = U_0 = E_P[Z_\tau | \mathcal{P}_0] \ .$$

But since $U_\tau \geq Z_\tau$ holds, it follows that

$$U_\tau = Z_\tau \ .$$

We now show that (U_t^τ) is a martingale. As already mentioned, from Doob's optional sampling theorem we know that (U_t^τ) is a supermartingale. Therefore:

$$
\begin{aligned}
U_0 \ &= \ U_0^\tau \geq E_P[U_t^\tau | \mathcal{P}_0] \\
&\geq \ E_P[\,E_P[U_{t+1}^\tau | \mathcal{P}_t]\,|\,\mathcal{P}_0\,] \\
&= \ E_P[U_{t+1}^\tau | \mathcal{P}_0] \\
&\geq \ E_P[\,E_P[U_T^\tau | \mathcal{P}_{t+1}]\,|\,\mathcal{P}_0\,] \\
&= \ E_P[U_T^\tau | \mathcal{P}_0] = E_P[U_\tau | \mathcal{P}_0] \\
&= \ E_P[Z_\tau | \mathcal{P}_0] \\
&= \ U_0 \ .
\end{aligned}
$$

From this we infer that

$$E_P[U_t^\tau | \mathcal{P}_0] = E_P[\,E_P[U_{t+1}^\tau | \mathcal{P}_t]\,|\,\mathcal{P}_0\,] \ . \tag{15.6}$$

But since $U_t^\tau \geq E_P[U_{t+1}^\tau | \mathcal{P}_t]$ it follows from (15.6) that

$$U_t^\tau = E_P[U_{t+1}^\tau | \mathcal{P}_t]$$

proving that (U_t^τ) is a martingale.   □

In view of this theorem it is natural to try to find optimal stopping times which relate to these two characterizing properties. We shall do so in the next two sections where we investigate the smallest stopping time for which the property $U_\tau = Z_\tau$ holds and the largest stopping time for which (U_t^τ) is a martingale.

15.3.3   Smallest Optimal Stopping Time

The smallest stopping time for which $U_\tau = Z_\tau$ holds is defined by

$$
\tau_{\min}(\omega) \overset{\text{def}}{=}
\begin{cases}
\min\{t; \ Z_t(\omega) = U_t(\omega)\} & \text{if } \{t; \ Z_t(\omega) = U_t(\omega)\} \neq \emptyset, \\
T & \text{otherwise} .
\end{cases}
$$

Since $\tau_{\min}$ is the first entry time of $W_t \overset{\text{def}}{=} Z_t - U_t$ into the set $A \overset{\text{def}}{=} \{0\}$, it is indeed a stopping time.

Proposition 15.14 *The stopping time $\tau_{\min}$ is the smallest optimal stopping time for (Z_t), i.e. if ν is another optimal stopping time, we have $\nu \geq \tau_{\min}$.*

Proof By the above proposition, to show optimality of $\tau \stackrel{\text{def}}{=} \tau_{\min}$ we just need to show that $Z_\tau = U_\tau$ and that (U_t^τ) is a martingale.

The first condition is an immediate consequence of the definition of τ. To show that (U_t^τ) is a martingale recall that

$$U_t^\tau = 1_{[\tau \geq t]} U_t + \sum_{s=0}^{t-1} 1_{[\tau = s]} U_s$$

holds. Hence,

$$E_P[U_{t+1}^\tau | \mathcal{P}_t] = 1_{[\tau \geq t+1]} E_P[U_{t+1} | \mathcal{P}_t] + \sum_{s=0}^{t} 1_{[\tau = s]} U_s , \qquad (15.7)$$

where we have used that for $0 \leq s \leq t$, the random variables $1_{[\tau = s]} U_s$ and $1_{[\tau \geq t+1]}$ are all $\mathcal{P}_t$-measurable.

On the set $[\tau \leq t] = [\tau \geq t+1]^c$ it is clear that

$$
\begin{aligned}
E_P[U_{t+1}^\tau | \mathcal{P}_t] &= 1_{[\tau \geq t+1]} E_P[U_{t+1} | \mathcal{P}_t] + \sum_{s=0}^{t} 1_{[\tau = s]} U_s \\
&= \sum_{s=0}^{t} 1_{[\tau = s]} U_s \\
&= 1_{[\tau = t]} U_t + \sum_{s=0}^{t-1} 1_{[\tau = s]} U_s \\
&= 1_{[\tau \geq t]} U_t + \sum_{s=0}^{t-1} 1_{[\tau = s]} U_s \\
&= U_t^\tau .
\end{aligned}
$$

On the other hand, by definition of τ we know that on the set $[\tau \geq t+1]$ we have

$$Z_t \neq U_t = \max\{Z_t, E_P[U_{t+1} | \mathcal{P}_t]\} .$$

Hence,

$$U_t = E_P[U_{t+1} | \mathcal{P}_t] .$$

Using this and (15.7) we therefore conclude that on this set

$$
\begin{aligned}
E_P[U_{t+1}^\tau | \mathcal{P}_t] &= 1_{[\tau \geq t+1]} U_t + \sum_{s=0}^{t} 1_{[\tau = s]} U_s \\
&= 1_{[\tau \geq t]} U_t + \sum_{s=0}^{t-1} 1_{[\tau = s]} U_s = U_t^\tau .
\end{aligned}
$$

This proves that (U_t^τ) is a martingale and, hence, the optimality of $\tau = \tau_{\min}$. That $\tau_{\min}$ is the smallest stopping time follows from the fact that, by definition, it is the smallest stopping time for which $U_\tau = Z_\tau$ and from the characterization of optimal stopping times.   □

15.3.4   Largest Optimal Stopping Time

In order to define the largest stopping time for which (U_t^τ) is a martingale we need the Doob decomposition of (U_t) :

$$U_t = M_t + A_t$$

where (M_t) is a martingale and (A_t) is a decreasing predictable process with $A_0 = 0$. Doob's decomposition was introduced in Chapter 10 .

Lemma 15.15 *Given a stopping time τ, the stopped process (U_t^τ) is a martingale if and only if $A_t^\tau = 0$.*

Proof First note that by representation (15.2)

$$U_t^\tau = M_t^\tau + A_t^\tau$$

holds. Since by the optional sampling theorem (M_t^τ) is also a martingale we see that (U_t^τ) is a martingale if and only if $A_t^\tau = 0$.   □

From this lemma it follows that the largest stopping time for which the stopped process (U_t^τ) is a martingale can be described by

$$\tau_{\max}(\omega) \stackrel{\text{def}}{=} \begin{cases} \min\{t;\ A_{t+1}(\omega) < 0\} & \text{if } \{t;\ A_{t+1}(\omega) < 0\} \neq \emptyset, \\ T & \text{otherwise .} \end{cases}$$

Note that $\tau_{\max}$ is indeed a stopping time since the process (A_t) is predictable.

Proposition 15.16 *$\tau_{\max}$ is the largest optimal stopping time, i.e. if ν is another optimal stopping time we have $\nu \leq \tau_{max}$.*

Proof Set $\tau \stackrel{\text{def}}{=} \tau_{\max}$. By our characterization of optimal stopping times we need only prove that $U_\tau = Z_\tau$. Fix an $\omega \in \Omega$. Then, $\omega \in [\tau = t]$ for some $t \in \{0, \ldots, T\}$. Hence, by definition of $\tau_{\max}$, we have for this t,

$$A_t(\omega) = 0 \qquad \text{and} \qquad A_{t+1}(\omega) < 0.$$

It follows that

$$U_t(\omega) = M_t(\omega) + A_t(\omega) = M_t(\omega)$$

and

$$
\begin{aligned}
E_P[U_{t+1}|\mathcal{P}_t](\omega) &= E_P[M_{t+1}|\mathcal{P}_t](\omega) + E_P[A_{t+1}(\omega)|\mathcal{P}_t] \\
&= M_t(\omega) + A_{t+1}(\omega) \\
&< M_t(\omega) = U_t(\omega),
\end{aligned}
$$

where we used that (M_s) is a martingale and that (A_s) is predictable. Hence,

$$U_\tau(\omega) = U_t(\omega) = \max\{Z_t(\omega), E_P[U_{t+1}|\mathcal{P}_t](\omega)\} = Z_t(\omega) = Z_\tau(\omega).$$

Since ω was arbitrary the proposition follows. $\qquad\square$

Remark 15.17 *Not every stopping time τ with $\tau_{\min} \leq \tau \leq \tau_{\max}$ is necessarily optimal.*

15.4 Markov Chains and the Snell Envelope

On a finite probability space (Ω, P) consider a stochastic process $(X_t)_{0 \leq t \leq T}$ with

$$X_t : \Omega \to E ,$$

where $E \subset \mathbb{R}$ is a finite set. The process (X_t) is called a *Markov chain* or *Markov process* if

$$P(X_{t+1} = a_{t+1} \mid X_0 = a_0, \ldots, X_t = a_t) = P(X_{t+1} = a_{t+1} \mid X_t = a_t) \qquad (15.8)$$

for any $a_0, \ldots, a_{t+1} \in E$ with $P(X_0 = a_0, \ldots, X_t = a_t) > 0$.
Note that the condition $P(X_0 = a_0, \ldots, X_t = a_t) > 0$ is imposed to guarantee that the conditional probabilities in (15.8) are well defined. If we interpret X_t as representing the state of a system at time t, then the Markov property requires that the probabilities of the possible states of X_{t+1} only depend on the state X_t and not on the previous states of the system from $t-1$ to 0. In other words, X_t has no memory.
In the next proposition we state a simple property of Markov chains which is an immediate consequence of identity (4.5) for conditional probabilities and of the Markov property (15.8).

Proposition 15.18 *Let (X_t) be a Markov chain on (Ω, P). Then, for any choice of $a_0, \ldots, a_t$ in E the following product formula holds:*

$$
\begin{aligned}
P(X_0 = a_0, &\ldots, X_t = a_t) \\
&= P(X_0 = a_0)\, P(X_1 = a_1 \mid X_0 = a_0) \ldots P(X_t = a_t \mid X_{t-1} = a_{t-1}).
\end{aligned}
$$

Examples 15.19 Consider a stochastic process (X_t), $t = 0, \ldots, T$ on (Ω, P) consisting of independent random variables. Then, the following assertions hold:

a) The process (X_t) is a Markov chain. Indeed, to see this just note that X_t is independent of $X_0, \ldots, X_{t-1}$. Therefore,

$$P(X_t = a_t | X_0 = a_0, \ldots, X_{t-1} = a_{t-1}) = P(X_t = a_t) .$$

The same argument yields

$$P(X_t = a_t | X_{t-1} = a_{t-1}) = P(X_t = a_t) ,$$

proving that (X_t) is Markov.

b) The process (S_t), where $S_t \stackrel{\text{def}}{=} \sum_{s=0}^{t} X_s$, is a Markov chain. To see this note that

$$P(S_t = a_t \quad | \quad S_0 = a_0, \ldots, S_{t-1} = a_{t-1})$$

$$= \frac{P(S_0 = a_0, \ldots, S_{t-1} = a_{t-1}, S_t = a_t)}{P(S_0 = a_0, \ldots, S_{t-1} = a_{t-1})}$$

$$= \frac{P(X_0 = a_0, X_1 = a_1 - a_0, \ldots, X_t = a_t - a_{t-1})}{P(X_0 = a_0, \ldots, X_{t-1} = a_{t-1} - a_{t-2})}$$

$$= P(X_t = a_t - a_{t-1}) = P(X_t = a_t - a_{t-1}) \frac{P(S_{t-1} = a_{t-1})}{P(S_{t-1} = a_{t-1})}$$

$$= \frac{P(X_t = a_t - a_{t-1}, S_{t-1} = a_{t-1})}{P(S_{t-1} = a_{t-1})} = \frac{P(S_t = a_t, S_{t-1} = a_{t-1})}{P(S_{t-1} = a_{t-1})}$$

$$= P(S_t = a_t | S_{t-1} = a_{t-1}) .$$

where we have used that X_t is independent of S_{t-1}.

Homogeneous Markov chains

A Markov chain (X_t) is called *homogeneous* if for any a, b in E the transition probabilities

$$P(X_{t+1} = b | X_t = a)$$

only depend on the values a, b but not on t. For a homogeneous Markov chain the *transition matrix* $P = [P(a, b)]_{a, b \in E}$ defined by

$$P(a, b) := P(X_{t+1} = b | X_t = a) ,$$

plays an important role. Note that the transition matrix P is a stochastic matrix, i.e.

$$P(a, b) \geq 0 \text{ and } \sum_{b \in E} P(a, b) = 1$$

for each $a \in E$.

Markov chains and their natural information structure

From now on we consider a homogeneous Markov chain (X_t), $t = 0, \ldots, T$ and its natural information structure $(\mathcal{P}_t)$. Recall that $\mathcal{P}_t$ was characterized by

$$\mathcal{P}_t = \{A_{t,\mathbf{a}}; \mathbf{a} = (a_1, \ldots, a_t) \in E^t\}$$

where

$$A_{t,\mathbf{a}} = \{\omega \in \Omega; X_1 = a_1, \ldots, X_t = a_t\} .$$

The following result says that some conditional expectations with respect to $\mathcal{P}_t$ depend only on the state of the Markov Chain at time t.

Proposition 15.20 *Let (X_t) be a homogeneous Markov chain. Then, for any function $f : \mathbb{R} \to \mathbb{R}$ we have*

$$E_P[f(X_{t+1}) \,|\, \mathcal{P}_t] = E_P[f(X_{t+1}) \,|\, X_t] .$$

More precisely, the conditional expectation of $f(X_{t+1})$ with respect to $\mathcal{P}_t$ can be expressed as

$$E_P[f(X_{t+1}) \,|\, \mathcal{P}_t](\omega) = \sum_{b \in E} P(b, X_t(\omega)) f(b) . \tag{15.9}$$

Proof We first look at the conditional expectation with respect to the atoms of $\mathcal{P}_t$. Since the Markov property implies that

$$
\begin{aligned}
P(X_{t+1} = b | A_{t,\mathbf{a}}) &= P(X_{t+1} = b | X_1 = a_1, \ldots, X_t = a_t) = P(X_{t+1} = b | X_t = a_t) \\
&= P(a_t, b) ,
\end{aligned}
$$

we infer that

$$E_P[f(X_{t+1}) \,|\, A_{t,\mathbf{a}}] = \sum_{b \in E} f(b) P(X_{t+1} = b | A_{t,\mathbf{a}}) = \sum_{b \in E} f(b) P(a_t, b).$$

Using this and going back to the definition of $E_P[f(X_{t+1}) \,|\, \mathcal{P}_t]$ in terms of the conditional expectations with respect to the atoms of the partition $\mathcal{P}_t$ we have

$$
\begin{aligned}
E_P[f(X_{t+1}) \,|\, \mathcal{P}_t](\omega) &= \sum_{\mathbf{a} \in E^t} E_P[f(X_{t+1}) \,|\, A_{t,\mathbf{a}}] 1_{A_{t,\mathbf{a}}}(\omega) \\
&= \sum_{a \in E} \sum_{\substack{\mathbf{a} \in E^t \\ a_t = a}} E_P[f(X_{t+1}) \,|\, A_{t,\mathbf{a}}] 1_{A_{t,\mathbf{a}}}(\omega) \\
&= \sum_{a \in E} \sum_{\substack{\mathbf{a} \in E^t \\ a_t = a}} [\sum_{b \in E} f(b) P(a, b)] 1_{A_{t,\mathbf{a}}}(\omega) \\
&= \sum_{a \in E} [\sum_{b \in E} f(b) P(a, b)] [\sum_{\substack{\mathbf{a} \in E^t \\ a_t = a}} 1_{A_{t,\mathbf{a}}}(\omega)] \\
&= \sum_{a \in E} [\sum_{b \in E} f(b) P(a, b)] 1_{X_t = a}(\omega) .
\end{aligned}
$$

Chapter 16

American Claims

A more common option, however, is one with exercise possible at any instant until the given future date. These options are termed American, and it is the added dimension which makes them more interesting and complex to evaluate.

R. Myneni

Up to now the focus of the book has been on the study of prices of European claims, i.e. of financial contracts which have a given payoff at a fixed maturity. We now turn to investigate financial contracts which allow for the possibility of early exercise. We will consider two issues which will actually turn out to be related to each other: pricing and hedging of American claims, on the one side, and optimal exercise strategies on the other.

16.1 The Underlying Economy

For the whole chapter we assume that a multi-period model, as described in Chapter 9, is given. In particular

- $\Omega = \{\omega_1, \omega_2, \ldots, \omega_n\}$ represents all possible states of the world;

- economic activity takes place at the dates $t = 0, 1, \ldots, T$;

- The information structure $\mathcal{I} = \{\mathcal{P}_0, \mathcal{P}_1, \ldots, \mathcal{P}_T\}$ describes the arrival of information as time elapses. Equivalently, as described in Chapter 7, we can also use the corresponding filtration $(\mathcal{A}_t)$, where $\mathcal{A}_t \stackrel{\text{def}}{=} \mathcal{A}(\mathcal{P}_t)$ is the algebra generated by $\mathcal{P}_t$;

- The probability measure P given by $p_i = P(\omega_i)$, describes the probabilities of each of the possible states of the world.

- $N + 1$ (non-dividend-paying) securities are traded. For $i = 0, 1, \ldots, N$ the price process of the i-th security, $(S_t^i)_{0 \le t \le T}$, is a $(\mathcal{P}_t)$-adapted and non-negative stochastic process.

- The 0-th security is a numeraire, i.e. $S_t^0(\omega) > 0$ for all $0 \le t \le T$ and $\omega \in \Omega$. This means that the prices of all other securities can always be discounted by the 0-th security, i.e. the process

$$\tilde{S}_t^j \stackrel{\text{def}}{=} \frac{S_t^j}{S_t^0}$$

  is well defined. Recall that discounting with respect to the 0-th security is nothing else than expressing the prices of other securities in terms of the price of the 0-th security.

We will further assume that our multi-period economy is both complete and free of arbitrage opportunities. Therefore, by the Fundamental Theorems of Asset Pricing of Chapter 11, there exists a unique probability measure Q on Ω under which the price processes of all the securities discounted by the 0-th security are martingales, i.e. such that

$$\tilde{S}_t^j = E_Q[\tilde{S}_{t+1}^j | \mathcal{P}_t]$$

holds for all j and t. Q was referred to as the risk-adjusted probability measure. In Chapter 11 we showed how to determine the fair value of an alternative X_S with maturity $S \le T$ as

$$\pi_t(X_S) = S_t^0 E_Q[\frac{X_S}{S_S^0} | \mathcal{P}_t]$$

for all $0 \le t \le S$.

Thus, if a stream of payments $\mathbf{X} = (X_t)$ is given, which entitles the holder at each time t to the payment X_t, we can consistently define its *fair value* as

$$\pi_t(\mathbf{X}) = \sum_{s=t}^{T} S_t^0 \cdot E_Q[\frac{X_s}{S_s^0} | \mathcal{P}_t]. \tag{16.1}$$

This formula will play an important role below.

16.2   American Claims Introduced

An American claim entitles the holder to early exercise. This means that the payoff obtained by the holder if he chooses to exercise at time $t \in \{1, \ldots, T\}$ needs to be specified. Denote the payoff random variable associated with early exercise at time t by $Z_t : \Omega \to \mathbb{R}$. We require that Z_t is $\mathcal{P}_t$-measurable. It follows that, mathematically, an American claim corresponds to a $(\mathcal{P}_t)$-adapted stochastic process $\mathbf{Z} = (Z_t)$ with $Z_t \ge 0$.

Note the difference from a stream of European claims $X_1, \ldots, X_T$ respectively maturing at $1, \ldots, T$. Such a stream entitles the holder to *all* payments $X_1, \ldots, X_T$. In contrast, an American claim entitles the holder to only *one* of the payments $Z_0, Z_1, \ldots, Z_T$. Which one of the payments the holder obtains depends on the particular time at which he chooses to exercise.

Examples 16.1   *a) An American call option on the j-th security with strike price K entitles the holder to buy, at any time prior to the maturity of the contract, a unit of the j-th security for K USD regardless of the prevailing price at the exercise time. Economically, this is equivalent to a cash settlement, i.e. to obtaining the payoff*

$$Z_t(\omega) := \max\{0, S_t^j(\omega) - K\} \, .$$

b) An American put option on the j-th security with strike price K entitles the holder to sell, at any time prior to the maturity of the contract, a unit of the j-th security for K USD regardless of the prevailing price at the exercise time. Economically, this is equivalent to obtaining a payoff of

$$Z_t(\omega) := \max\{0, K - S_t^j(\omega)\} \, .$$

c) Let $X : \Omega \to \mathbb{R}^+$ represent the payoff of a European claim maturing at time T. Then,

$$Z_t := \begin{cases} 0 & t < T \, , \\ X & t = T \, , \end{cases}$$

is a $(\mathcal{P}_t)$-adapted process. Hence, formally, European claims can also be considered as a special case of American claims.

Examples 16.1 a) and b) are typical for the structure of many American claims found in applications: the payoff is a function of the price of the underlying basic security and the holder can choose the time at which to obtain it. American claims are more complex to value than their European counterparts because one needs to determine whether there is any value in the ability to exercise earlier. Whether this is the case or not depends on the type of claim. Later we will see that while for an American call it never pays to exercise before maturity, there are circumstances where the early exercise of an American put is optimal.

Remark 16.2 *We only consider American options with non-zero payoffs at times $1, \ldots, T$. For notational ease, especially in backward recursion formulas, we will sometimes use $Z_0 \overset{\text{def}}{=} 0$.*

16.3   The Buyer's Perspective: Optimal Exercise

One of the central questions facing the buyer or holder of an American claim $\mathbf{Z} = (Z_t)$ is which strategy to pursue when deciding when to exercise his right. At a given time t_0 the holder has the option to exercise at any time $t \geq t_0$. An exercise strategy at time t_0 corresponds to a rule which tells the holder at any time $t \geq t_0$ whether to exercise or not. Obviously, the holder should be able to decide this on the grounds of the information $\mathcal{P}_t$ which is available at time t. Thus, an *exercise strategy at time* t_0 corresponds to a stopping time $\tau \in \mathcal{S}_{t_0,T}$, i.e.

$$\tau : \Omega \to \{t_0, t_0 + 1, \dots, T\} \ .$$

We now describe the payoff of an American claim if at time t_0 the holder decides to pursue an exercise strategy described by the stopping time $\tau \in \mathcal{S}_{t_0,T}$. Given $\omega \in \Omega$ the holder exercises the option at time $\tau(\omega)$ and thus receives the payoff $Z_\tau(\omega) = Z_{\tau(\omega)}(\omega)$. This means that for any $t \in \{t_0, t_0+1, \dots, T\}$ the payoff $X_t^{(\mathbf{Z},\tau)}$ generated by the American claim $\mathbf{Z}$ under the exercise strategy τ will be given by

$$X_t^{(\mathbf{Z},\tau)} \stackrel{\text{def}}{=} Z_t \, 1_{[\tau=t]} \ .$$

Note that $\mathbf{X}^{(\mathbf{Z},\tau)} \stackrel{\text{def}}{=} (X_t^{(\mathbf{Z},\tau)})$ represents a stream of European claims, where $X_t^{(\mathbf{Z},\tau)}$ is the European claim maturing at time t. By the results of Chapters 9 and 11, its fair value is thus given by

$$\pi_{t_0}(\mathbf{X}^{(\mathbf{Z},\tau)}) \ = \ \sum_{t=t_0}^{T} E_Q[X_t^{(\mathbf{Z},\tau)} \frac{S_{t_0}^0}{S_t^0} | \mathcal{P}_{t_0}]$$

$$= \ \sum_{t=t_0}^{T} E_Q[Z_t \, 1_{[\tau=t]} \frac{S_{t_0}^0}{S_t^0} | \mathcal{P}_{t_0}] = E_Q[Z_\tau \frac{S_{t_0}^0}{S_\tau^0} | \mathcal{P}_{t_0}] \ .$$

We call this the *value of the American claim when exercising according to* τ.

Optimal exercise

From the point of view of the holder of the American claim $\mathbf{Z} = (Z_t)$, it is interesting to know which exercise strategies are optimal. We have just determined the value of the American claim when exercising according to a given strategy. Therefore, it is clear that at time t_0 an exercise strategy $\tau \in \mathcal{S}_{t_0,T}$ will be *optimal* if the value of the American claim when exercising according to τ is larger than or equal to the value of the claim when exercising according to any other strategy, i.e. if

$$E_Q[Z_\tau \frac{S_{t_0}^0}{S_\tau^0} | \mathcal{P}_{t_0}] = \max_{\mu \in \mathcal{S}_{t_0,T}} E_Q[Z_\mu \frac{S_{t_0}^0}{S_\mu^0} | \mathcal{P}_{t_0}] \ .$$

Note that this maximum indeed exists since the set $\mathcal{S}_{t_0,T}$ of stopping times is finite. Since the buyer can choose an optimal strategy τ he should be prepared to pay the amount $E_Q[Z_\tau \frac{S_{t_0}^0}{S_\tau^0}|\mathcal{P}_{t_0}]$ for the claim. On the other hand any other strategy will result in a payoff with a lower value so that he should not be willing to pay more than this amount. For this reason we call

$$\max_{\mu \in \mathcal{S}_{t_0,T}} E_Q[Z_\mu \frac{S_{t_0}^0}{S_\mu^0}|\mathcal{P}_{t_0}]$$

the *buyer's price*.

Observe that the optimality of τ is equivalent to

$$E_Q[\tilde{Z}_\tau|\mathcal{P}_{t_0}] = \tilde{U}_{t_0} \overset{\text{def}}{=} \max_{\mu \in \mathcal{S}_{t_0,T}} E_Q[\tilde{Z}_\mu|\mathcal{P}_{t_0}] \tag{16.2}$$

where $\tilde{\mathbf{Z}} = (\tilde{Z}_t)$ is defined by

$$\tilde{Z}_t \overset{\text{def}}{=} \frac{Z_t}{S_t^0} .$$

Recall that the process $(\tilde{U}_t)$ was called the Snell envelope of $(\tilde{Z}_t)$ and was the smallest Q-supermartingale dominating $(\tilde{Z}_t)$. Moreover, $(\tilde{U}_t)$ was also characterized by the backward recursion relation

$$\tilde{U}_t = \max\{\tilde{Z}_t, E_Q[\tilde{U}_{t+1}|\mathcal{P}_t]\} .$$

16.4 The Seller's Perspective: Hedging

From the perspective of the seller or writer it is not optimal exercise which is the main issue, but optimal hedging. The writer of an American claim $\mathbf{Z} = (Z_t)$ needs to make sure that at any time he will be able to honor his obligation should the holder choose to exercise then.

A *hedge* or *hedging strategy* for $\mathbf{Z}$ at time t_0 is a self-financing portfolio $\Phi = (\phi_t)_{t \geq t_0}$ such that

$$V_t(\Phi) \geq Z_t \tag{16.3}$$

holds for all $t \geq t_0$. We shall sometimes use the term *time-t_0 hedge* for $\mathbf{Z}$ instead of the more cumbersome *hedge at time t_0*.

If Φ is a time-t_0 hedge for $\mathbf{Z}$ and we sell $\mathbf{Z}$ for the amount $V_{t_0}(\Phi)$, with the proceeds we can implement the hedging strategy Φ. Inequality (16.3) tells us that at any time $t \geq t_0$ we will have enough wealth to pay out Z_t should the holder of the claim choose to exercise then.

Constructing hedges from martingales

If $\Phi = (\phi_t)_{t \geq t_0}$ is a time-t_0 hedge for $\mathbf{Z}$ we know that $(\tilde{V}_t(\Phi))_{t \geq t_0}$ is a Q-martingale which dominates $(\tilde{Z}_t)_{t \geq t_0}$. In fact there is a one-to-one correspondence between Q-martingales $(N_t)_{t \geq t_0}$ dominating $(\tilde{Z}_t)_{t \geq t_0}$ and time-t_0 hedges for $\mathbf{Z}$. To see this let $(N_t)_{t \geq t_0}$ be a Q-martingale dominating $(\tilde{Z}_t)_{t \geq t_0}$ and set

$$X \stackrel{\text{def}}{=} N_T \cdot S_T^0 \ .$$

Thus, X is a European claim which by the completeness of the market can be replicated by a self-financing strategy $\Phi = (\phi_t)$. In other words we have

$$V_T(\Phi) = X \ .$$

Moreover, we also know that $(\tilde{V}_t(\Phi))$ is a Q-martingale with the same terminal value as $(N_t)_{t \geq t_0}$. We thus conclude from Proposition 10.12 that $(\tilde{V}_t(\Phi))_{t \geq t_0}$ and $(N_t)_{t \geq t_0}$ coincide. Since $(N_t)_{t \geq t_0}$ dominates $(\tilde{Z}_t)_{t \geq t_0}$ we infer that

$$V_t(\Phi) \geq Z_t$$

holds for all $t \geq t_0$. We have thus constructed a time-t_0 hedge Φ with value process equal to $(N_t \cdot S_t^0)_{t \geq t_0}$.

Optimal hedging

From the point of view of the writer optimality presents itself from a different side. He will look at the optimal way of hedging a given American claim $\mathbf{Z} = (Z_t)$. The result of the previous section suggests how to look for a cheapest time-t_0 hedge. Recall that the Snell envelope $(\tilde{U}_t)_{t \geq t_0}$ of $(\tilde{Z}_t)_{t \geq t_0}$ was shown to be the smallest Q-supermartingale dominating $(\tilde{Z}_t)_{t \geq t_0}$. The cheapest hedge will be obtained by applying the above construction principle to the martingale part of the Doob decomposition of the Snell envelope.
In order to do that let

$$\tilde{U}_t = \tilde{M}_t^{t_0} + \tilde{A}_t^{t_0}$$

be the Doob decomposition of $(\tilde{Z}_t)_{t \geq t_0}$, i.e. $(\tilde{M}_t^{t_0})_{t \geq t_0}$ is a Q-martingale and $(\tilde{A}_t^{t_0})_{t \geq t_0}$ is predictable and decreasing with $A_{t_0} = 0$.

Remark 16.3 *It is important to note that while the Snell envelope $(U_t)_{t \geq t_0}$ does not depend on the starting point t_0, its Doob decomposition — and, with it, also $(\tilde{M}_t^{t_0})_{t \geq t_0}$ — does.*

Let Φ^* be the hedge corresponding to the Q-martingale $(\tilde{M}_t^{t_0})$. The next result shows that Φ^* is a time-t_0 cheapest hedge for $(Z_t)_{t \geq t_0}$.

Theorem 16.4 *The hedging strategy Φ^* constructed above is a cheapest hedge, i.e.*

$$V_{t_0}(\Phi^*) = \inf_{\Phi}\{V_0(\Phi)\,;\, \Phi \text{ is a time-}t_0 \text{ hedge for } \mathbf{Z}\}.$$

Moreover,

$$V_{t_0}(\Phi^*) = S_{t_0}^0 \tilde{U}_{t_0}$$

where $(\tilde{U}_t)_{t \geq t_0}$ is the Snell envelope of $(\tilde{Z}_t)_{t \geq t_0}$, which is given by the recursion formula

$$\begin{aligned}
\tilde{U}_T &= \tilde{Z}_T\,, \\
\tilde{U}_t &= \max\{\tilde{Z}_t\,,\, E_Q[\tilde{U}_{t+1}|\mathcal{P}_t]\}\,.
\end{aligned}$$

Proof Let $\tau \in \mathcal{S}_{t_0,T}$ be optimal stopping time. Then, by construction of Φ^* and equation (16.2) we have

$$V_{t_0}[\Phi^*] = S_{t_0}^0 \tilde{M}_{t_0}^{t_0} = S_{t_0}^0 \tilde{U}_{t_0} = E_Q[Z_\tau \frac{S_{t_0}^0}{S_\tau^0}]\,. \tag{16.4}$$

Let Ψ be any hedge for $(Z_t)_{t \geq t_0}$. Setting $N_t \overset{\text{def}}{=} V_t[\Psi]$ we have

$$N_t \geq Z_t\,.$$

In particular, for any stopping time $\tau \in \mathcal{S}_{t_0,T}$ we obtain

$$N_\tau \geq Z_\tau\,.$$

Hence, using (16.4), the facts that $(\frac{N_t}{S_t^0})$ is a Q-martingale, and that stopped martingales remain martingales (Theorem 15.6 and Proposition 15.7) we find

$$\begin{aligned}
V_{t_0}[\Phi^*] &= E_Q[Z_\tau \frac{S_{t_0}^0}{S_\tau^0}|\mathcal{P}_{t_0}] \\
&\leq E_Q[N_\tau \frac{S_{t_0}^0}{S_\tau^0}|\mathcal{P}_{t_0}] = S_{t_0}^0 E_Q[\frac{N_\tau}{S_\tau^0}|\mathcal{P}_{t_0}] = S_{t_0}^0 \frac{N_{t_0}}{S_{t_0}} = N_{t_0} \\
&= V_{t_0}[\Psi]\,.
\end{aligned}$$

$\square$

The theorem shows that in a complete and arbitrage-free market every American claim $\mathbf{Z} = (Z_t)$ admits at any time t_0 a cheapest hedge. We note here that a cheapest hedge need not be unique.

Proof Recall that if $\tilde{U}_t$ is the Snell envelope of $(\tilde{Z}_t)$ then,

$$\pi_t(\mathbf{Z}) = S_{t_0}^0 \cdot \tilde{U}_{t_0}$$

holds. First, assume that $\pi_t(\mathbf{Z}) = \pi_t(X)$ holds. Then,

$$\tilde{\pi}_t(X) = \tilde{\pi}_t(\mathbf{Z}) = \tilde{U}_t .$$

Since $(\tilde{U}_t)$ dominates $(\tilde{Z}_t)$ we infer that $(\tilde{\pi}_t(X))$ dominates $(\tilde{Z}_t)$, which is equivalent to $\pi_t(X) \geq Z_t$ for all $t \geq t_0$.

Assume now that $\pi_t(X) \geq Z_t$. It follows that $(\tilde{\pi}_t(X))$ is a martingale dominating $(\tilde{Z}_t)$. Since the Snell envelope $(\tilde{U}_t)$ of $(\tilde{Z}_t)$ is the smallest supermartingale dominating $(\tilde{Z}_t)$, we infer that $\tilde{\pi}_t(\mathbf{Z}) = \tilde{U}_t \leq \tilde{\pi}_t(X)$. This is equivalent to $\pi_t(\mathbf{Z}) \leq \pi_t(X)$, which together with (16.6) implies $\pi_t(\mathbf{Z}) = \pi_t(X)$. $\square$

16.7   Homogeneous Markov Processes

The value process (U_t) of an American claim $\mathbf{Z} = (Z_t)$ was shown to be given by the recursion formula (16.5), namely

$$U_T = Z_T ,$$

$$U_t = \max\{Z_t , E_Q[U_{t+1} \frac{S_t^0}{S_{t+1}^0}|\mathcal{P}_t]\} .$$

In order to evaluate this expression we need to know the history of the economy up to time t since a conditional expectation with respect to $\mathcal{P}_t$ needs to be taken. For the class of homogeneous Markov processes, however, it is possible to show that it suffices to know the state of the economy at the time of valuation.

Consider a market where the 0-th security (the numeraire) is deterministic and the price process of one of the other securities, denoted by (S_t) is a homogeneous Markov process. The type of American claim we will deal with are of the form $\mathbf{Z} = (Z_t)$, where

$$Z_t = f_t(S_t) ,$$

with arbitrary functions $f_t : \mathbb{R} \to \mathbb{R}$. For these American claims we have the following result.

Theorem 16.6 *For* $\mathbf{Z} = (Z_t)$ *given as above, the fair value is given by*

$$\pi_t(\mathbf{Z}) = h_t(S_t) ,$$

where the functions $h_t : \mathbb{R} \to \mathbb{R}$ *are defined recursively by*

$$h_T(s) = f_T(s) ,$$

$$h_t(s) = \max\{f_t(s) , E_Q[h_{t+1}(S_{t+1}) \frac{S_t^0}{S_{t+1}^0}|S_t = s]\} .$$

Proof By Theorem 16.4 and the definition of fair value we know that

$$\pi_t(\mathbf{Z}) = S_t^0 \, \tilde{U}_t$$

where $(\tilde{U}_s)$ is the Snell envelope of $(\tilde{Z}_t)$. Observing that $\tilde{Z}_t = \tilde{f}_t(S_t)$ with

$$\tilde{f}(s) \stackrel{\text{def}}{=} \frac{f_t(s)}{S_t^0} \, ,$$

we can apply the results on the Snell envelope of functions of Markov processes (Theorem 15.21) to obtain

$$\tilde{U}_t = \tilde{h}_t(S_t) \, ,$$

where $\tilde{h}$ is defined recursively

$$\begin{aligned}
\tilde{h}_T(s) &= \tilde{f}_T(s) \, , \\
\tilde{h}_t(s) &= \max\{\tilde{f}_t(s) \, , \sum_{b \in \mathbb{R}} Q(S_{t+1} = b | S_t = s) \tilde{h}_{t+1}(b)\} \\
&= \max\{\tilde{f}_t(s) \, , \, E_Q[\tilde{h}_{t+1}(S_{t+1}) | S_t = s]\} \, .
\end{aligned}$$

Setting $h_t = S_t^0 \tilde{h}_t$ immediately yields

$$\pi_t(\mathbf{Z}) = S_t^0 \tilde{U}_t = \max\{f_t(s) \, , \, E_Q[h_{t+1}(S_{t+1}) \frac{S_t^0}{S_{t+1}^0} \, | \, S_t = s]\} \, .$$

$\square$

16.7.1   Cox–Ross–Rubinstein Setting

We now apply the results on Markov processes to the Cox–Ross–Rubinstein model described in Chapter 12. During each of the N periods, the N-period economy considered in that chapter can have a "good" or a "bad" development with probabilities p and $1 - p$, respectively. Below we recall some facts and notation. The underlying sample space is

$$\Omega = \{0, 1\}^N \, .$$

An elementary event $\omega = (\omega_1, \ldots, \omega_N) \in \Omega$ describes a possible trajectory of the economy where $\omega_t = 1$ means that the development was "good" during the period $[t-1, t]$. Correspondingly $\omega_t = 0$ represents a "bad" development during the period $[t - 1, t]$.

Thus, the random variable $Z_t : \Omega \to \mathbb{R}$ given by

$$Z_t(\omega) \overset{\text{def}}{=} \omega_t$$

for $\omega \in \Omega$ detects whether the development during the period $[t-1, t]$ was "good" or "bad" and, consequently, the random variable $D_t : \Omega \to \mathbb{R}$ given by

$$D_t \overset{\text{def}}{=} \sum_{s=1}^{t} Z_s$$

counts the number of periods up to time t under which the development of the economy was "good".

The probability of an elementary event is given by

$$P(\omega) = p^{D_N(\omega)}(1-p)^{N-D_N(\omega)} .$$

The random variables $Z_1, \dots, Z_N$ are independent with respect to P.

In the Cox–Ross–Rubinstein economy two securities are traded. First, a risk-free instrument, the money-market account, growing deterministically at the risk-free rate r. Second, a risky instrument, the stock, which at each period can have either a return of y_g if the economy has a "good" development or a return of y_b if the economy has a "bad" development.

To ensure that the model is free of arbitrage opportunities and complete, we will assume throughout (see Corollary 12.8) that

$$-1 < y_b < r < y_g . \tag{16.7}$$

The stock price process is thus described by

$$S_t(\omega) \overset{\text{def}}{=} S_0(1 + y_g)^{D_t(\omega)}(1 + y_b)^{t - D_t(\omega)}$$

where S_0 denotes the initial price. The money-market account is described by

$$B_t(\omega) \overset{\text{def}}{=} (1 + r)^t .$$

The risk-adjusted probability measure was shown to be given by

$$Q(\omega) = q^{D_N(\omega)}(1-q)^{N-D_N(\omega)} ,$$

where

$$q \overset{\text{def}}{=} \frac{r - y_b}{y_g - y_b} .$$

It is important to note that $Z_1, \dots, Z_N$ are independent also under this new probability measure (see Remark 12.5). A consequence of this is the fact that (S_t) is a homogeneous Markov process with respect to the risk-adjusted probability measure.

Lemma 16.7 *For any choice of* $a_0, \ldots, a_{t+1}$ *satisfying*

$$Q(S_0 = a_0, \ldots, S_t = a_t) > 0$$

we have

$$Q(S_{t+1} = a_{t+1} | S_0 = a_0, \ldots, S_t = a_t) = Q(S_{t+1} = a_{t+1} | S_t = a_t) .$$

Proof The independence of $Z_1, \ldots, Z_n$ together with Example 15.19 implies that (D_t) is a Markov process with respect to Q.

To establish that (S_t) is Markov, first note that

$$D_t = g(S_t) \stackrel{\text{def}}{=} \log[\frac{S_t}{S_0(1+y_b)}]/\log[\frac{S_0(1+y_g)}{S_0(1+y_b)}]$$

so that

$$
\begin{aligned}
& Q(S_{t+1} = a_{t+1} | S_0 = a_0, \ldots, S_t = a_t) \\
= \ & Q(D_{t+1} = g(a_{t+1}) | D_0 = g(a_0), \ldots, D_t = g(a_t)) \\
= \ & Q(D_{t+1} = g(a_{t+1}) | D_t = g(a_t)) \\
= \ & Q(S_{t+1} = a_{t+1} | S_t = a_t) ,
\end{aligned}
$$

where we have used that (D_t) is Markov. It follows that (S_t) is Markov. Moreover, (S_t) is homogeneous, i.e. $Q(S_{t+1} = b | S_t = a)$ is independent of t, since

$$
\begin{aligned}
Q(S_{t+1} = b | S_t = a) \ & = \ Q(D_{t+1} = g(b) | D_t = g(a)) \\
& = \ \frac{Q(D_{t+1} = g(b), D_t = g(a))}{Q(D_t = g(a))} \\
& = \ \frac{Q(Z_{t+1} = g(b) - g(a), D_t = g(a))}{Q(D_t = g(a))} \\
& = \ \frac{Q(Z_{t+1} = g(b) - g(a)) \, Q(D_t = g(a))}{Q(D_t = g(a))} \\
& = \ Q(Z_{t+1} = g(b) - g(a)) ,
\end{aligned}
$$

where we have used the fact that Z_{t+1} is independent of D_t. $\qquad\square$

An immediate application of Theorem 16.6 yields the following result.

Proposition 16.8 *If the American claim* (Z_t) *is given by* $Z_t \stackrel{\text{def}}{=} f_t(S_t)$ *for some functions* $f_t : \mathbb{R} \to \mathbb{R}$, *then its value at time* t *depends only on the value of* S_t *and not on the previous values* $S_0, \ldots, S_{t-1}$. *More precisely, for the value* $h_t(s)$ *of the claim, given that we know that* $S_t = s$, *we have the backward recursive formulas*

$$
\begin{aligned}
h_T(s) \ & = \ f_T(s) \\
h_t(s) \ & = \ \max\{f_t(s), \frac{q \cdot h_{t+1}(s(1+y_g)) + (1-q) \cdot h_{t+1}(s(1+y_b))}{1+r}\} .
\end{aligned}
$$

Dependence on the price of the underlying

The above recursive formula can be used to see that many structural properties of the payoff functions f_t, $1 \le t \le T$ — such as continuity, monotonicity or convexity — are inherited by the value functions h_t.

Recall that a function $g : I \to \mathbb{R}$, where I is an interval in $\mathbb{R}$, said to be *non-increasing* if

$$s_1 < s_2 \Rightarrow g(s_1) \ge g(s_2) ,$$

and *convex* if for all $s_1, s_2 \in I$ and $\lambda \in [0, 1]$ we have

$$g(\lambda s_1 + (1 - \lambda)s_2) \le \lambda g(s_1) + (1 - \lambda)g(s_2) .$$

Proposition 16.9 *Assume that for all $1 \le t \le T$ the payoff functions f_t are either continuous, decreasing or convex. Then, the same is true for the value functions h_t.*

Proof For any function $g : \mathbb{R}_+ \to \mathbb{R}$ define the function $\hat{g} : \mathbb{R}_+ \to \mathbb{R}$ by

$$\hat{g}(s) \stackrel{\text{def}}{=} \frac{q \cdot g(s(1 + y_g)) + (1 - q) \cdot g(s(1 + y_b))}{1 + r}$$

It is easy to see that if g is either continuous, decreasing or convex, then the same is true of $\hat{g}$. Here, for the preservation of monotonicity, it is important that, by assumption, $1 + y_b$ and $1 + y_g$ are both positive.

Using this observation it is easy to prove the proposition by using the recursive formulas of Proposition 16.8 and a backward induction argument.

Assume first that the payoff functions f_t are convex. Since $h_T(s) = f_T(s)$ holds, it follows that h_T is trivially also convex.

Suppose now that we have shown that h_t is convex for some t. We show that h_{t-1} is convex as well. From Proposition 16.8 we have

$$h_{t-1}(s) = \max\{f_{t-1}(s), \hat{h}_t(s)\} .$$

As we have just observed above $\hat{h}_t$ inherits the convexity property from h_t. In addition, from Proposition A.9, we know that taking the maximum of two convex functions preserves convexity. Therefore, we conclude that h_{t-1} is convex as claimed.

To prove the statement in the case of continuous, respectively decreasing, payoff functions we just need to replace the word "convex" by the word "continuous" or "decreasing", respectively. In the case of continuity we also need to invoke Proposition A.8 instead of A.9.   □

Pricing American puts

Recall that the payoff of an American put is described by

$$Z_t(\omega) := (K - S_t(\omega))_+ \, ,$$

where for convenience we will use the notation

$$(x)_+ \stackrel{\text{def}}{=} \max\{0, x\} \, .$$

In Section 16.6 an American call was shown to have the same value as a European call, because for an American call, early exercise is always suboptimal. This is not the case with American put options since early exercise can be shown to be optimal in certain instances.

To get an intuition for this phenomenon assume that $r > 0$ and that at a given time t_0 the price s of the underlying stock satisfies

$$0 \leq s \leq \frac{r}{1+r} K \, . \tag{16.8}$$

In this case, exercising the put at this point yields the payoff $K - s$ which satisfies

$$K - s \geq \frac{1}{1+r} K \, .$$

Thus, investing the proceeds in the risk free security, the holder will have at each time $t > t_0$ n investment worth $(K - s)(1+r)^{t-t_0}$. By (16.8) we get

$$(K - s)(1+r)^{t-t_0} > K(1+r)^{t-1-t_0} > K \, .$$

Now, since the payoff of the option at any time is at most K, it follows that the holder will be better off by exercising the put and investing in the risk-free security than by holding on to the option.

To be able to to give a more complete picture of when it pays to exercise early — and because it is of interest and useful on its own right — we note that an immediate application of Proposition 16.8 yields a recursive pricing formula for the American put with strike price K :

$$
\begin{aligned}
P_T(s) &= (K - s)_+ \\
P_t(s) &= \max\Big\{(K - s)_+, \frac{q \cdot P_{t+1}(s(1 + y_g)) + (1 - q) \cdot P_{t+1}(s(1 + y_b))}{1 + r}\Big\} \, ,
\end{aligned}
$$

where $P_t(s)$ denotes the price of the put at time t given that the observed price of the stock is $S_t = s$.

Noting that the (time-independent) payoff function of the put, given by

$$f(x) \stackrel{\text{def}}{=} \max\{0, K - x\} \, ,$$

is continuous, decreasing and convex we can immediately conclude from Proposition 16.9 that the price of a put at any given time is a continuous, decreasing and convex function of the price of the underlying stock.

Proposition 16.10 *Let t_0 be fixed. The function $s \mapsto P_{t_0}(s)$ is continuous, decreasing and convex.*

We can now start giving a more complete description of the dependency of the value of an American put on the price of the underlying. We start by showing that in the Cox–Ross–Rubinstein, if the price is high enough, the put has no value. This is because in such a model if the price is too high there is no possibility for the price of the underlying to fall below the strike price of the option.

Lemma 16.11 *For $s \in [\frac{K}{(1+y_b)^{T-t_0}}, \infty)$ we have $P_{t_0} = 0$. Conversely, if $s \in [0, \frac{K}{(1+y_b)^{T-t_0}})$ we have $P_{t_0} > 0$.*

Proof We will use the observation that if at time t_0 the price of the underlying is such that for all possible developments of the economy in the subsequent periods the payoff of the put is zero, then the value of the put at time t_0 must also be zero. Conversely, if there exists one possible development of the economy in the subsequent periods for which the payoff is strictly positive, then the value of the put at time t_0 must also be strictly positive.

Assume that $s \in [\frac{K}{(1+y_b)^{T-t_0}}, \infty)$. Then, the price of the underlying will always remain above $s(1 + y_b)^{T-t_0}$. By assumption we have

$$s(1 + y_b)^{T-t_0} \geq K \ ,$$

so that for the the payoff $(S_t - K)_+$ at any time $t_0 \leq t \leq T$ we will always have $(S_t - K)_+ = 0$. By our introductory remark we conclude that $P_{t_0}(s) = 0$.

Assume now that $s \in [0, \frac{K}{(1+y_b)^{T-t_0}})$. Then, if at each period the economy has a "bad" development, the price of the underlying at time T will be $s(1 + y_b)^{T-t_0}$. By assumption we have

$$s(1 + y_b)^{T-t_0} < K \ ,$$

so that at least for this particular trajectory we will have that the payoff $(S_t - K)_+$ at time T we will have

$$(S_t - K)_+ = s(1 + y_b)^{T-t_0} - K > 0 \ .$$

By our introductory remark we conclude that $P_{t_0}(s) > 0$. $\square$

As a next step we show that early exercise of an American put is always optimal as long as the price of the underlying is low enough.

Lemma 16.12 *Let t_0 be fixed. Then, there exists a $\delta_{t_0} > 0$ such that for $0 \leq s < \delta_{t_0}$, we always have*

$$P_{t_0}(s) = K - s \ .$$

Proof We prove this by backward induction. For $t_0 = T$ the assertion is obvious since then $P_T(s) = (K - s)_+$. Assume we had showed that there exists $\delta_t > 0$ such that

$$P_t(s) = K - s \qquad (16.9)$$

for $0 \le s < \delta_t$ and set

$$\delta_{t-1} \stackrel{\text{def}}{=} \frac{\delta_t}{1 + y_g} > 0 .$$

Then, for $0 \le s < \delta_{t-1}$ we have

$$s(1 + y_b) < s(1 + y_g) < \delta_t$$

Therefore, by assumption (16.9), for $0 \le s < \delta_t$ we have

$$\frac{1}{1 + r}[q \cdot P_t(s(1 + y_g)) + (1 - q) \cdot P_t(s(1 + y_b))]$$

$$= \frac{1}{1 + r}[\frac{r - y_b}{y_g - y_b} \cdot (K - s(1 + y_g)) + \frac{y_g - r}{y_g - y_b} \cdot (K - s(1 + y_b))]$$

$$= \frac{1}{1 + r}[K - s - r]$$

$$\le K - s$$

Using the recursion formula, it follows that for $0 \le s < \delta_t$ we have

$$
\begin{aligned}
P_{t-1}(s) &= \max\{(K - s)_+, \frac{q \cdot P_t(s(1 + y_g)) + (1 - q) \cdot P_t(s(1 + y_b))}{1 + r}\} \\
&= K - s .
\end{aligned}
$$

This completes the proof of the lemma. □

Remark 16.13 *It is important to be aware that while the statement of the above lemma is also true for the continuous-time setting if $r > 0$, it no longer holds for $r = 0$ (see [45], Section 8.3).*

As a consequence we obtain a neat result which provides a complete qualitative description of the stock price dependency of the value of an American put. It shows that at any time there is a range of values for which it pays to exercise.

Theorem 16.14 *Let t_0 be fixed and assume that $y_b < 0$. Then, there exists an $s^* \in (0, K)$ such that*

$$P_{t_0}(s) = K - s \qquad \text{if } s \in [0, s^*] ,$$

$$P_{t_0}(s) > K - s \qquad \text{if } s \in (s^*, \tfrac{K}{(1+y_b)^{T-t_0}}) , \text{ and}$$

$$P_{t_0}(s) = 0 \qquad \text{if } s \in [\tfrac{K}{(1+y_b)^{T-t_0}}, \infty) .$$

Proof Set

$$s^* \stackrel{\text{def}}{=} \sup\{s \geq 0 ; \ P_{t_0}(s) = K - s\}$$

By Lemma 16.12 we have that $s^* > 0$. Furthermore, for $s \geq K < \frac{K}{(1+y_b)^{T-t_0}}$ we have by Step II that

$$K - s \leq 0 < P_{t_0}(s) .$$

Therefore, $s^* < K$.

We now show that $P_{t_0}(s^*) = K - s^*$. Indeed, take a sequence (s_n) converging to s^* and satisfying $P_{t_0}(s_n) = K - s_n$, which is possible by definition of s^*. Then, by the continuity of P_{t_0} we conclude

$$P_{t_0}(s^*) = \lim_{n \to \infty} P_{t_0}(s_n) = \lim_{n \to \infty} (K - s_n) = K - s^* .$$

Finally, we show that for all $s \in [0, s^*]$ we have $P_{t_0} = K - s$. Indeed, we can write $s = \lambda s^*$ with $\lambda \in [0, 1]$. By the convexity of P_{t_0} we obtain

$$
\begin{aligned}
P_{t_0}(s) &= P_{t_0}(\lambda s^* + (1 - \lambda)0) \\
&\leq \lambda P_{t_0}(s^*) + (1 - \lambda)P_{t_0}(0) \\
&= \lambda(K - s^*) + (1 - \lambda)K \\
&= K - \lambda s^* \\
&= K - s .
\end{aligned}
$$

By the recursion formula we always have $P_{t_0}(s) \geq K - s$ so that we can infer that $P_{t_0} = K - s$. This concludes the proof of the theorem.   □

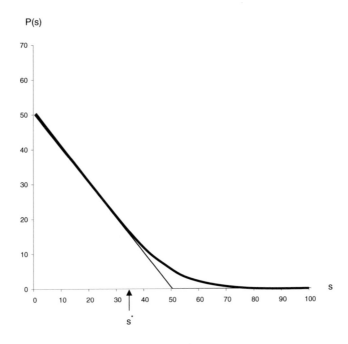

Figure 16.1: Payoff function of an American put.

Concluding Remarks and Suggested Reading

With the study of American claims we have completed the study of contingent
claims pricing in finite probability space. We should note, however, that we have
not considered the important case where the underlying stock pays dividends. This
feature is not very difficult to incorporate and a treatment can be found in [31].
Some of the results do change however. In particular, the price of an American
and a European call no longer coincide, i.e. early exercise can be optimal in certain
circumstances.

In Chapter 14 we saw in the case of European options that an analytical option
pricing formula in continuous time can be obtained by studying the limiting behavior of the Cox–Ross–Rubinstein model as the length of the time intervals tends
to zero. No corresponding formula is known for American options. The limiting
behavior, however, can still be studied (see [40] and the references cited therein).
Other treatments can be found in [39], [47] and [45].

Appendix A

Euclidean Space and Linear Algebra

In this appendix we give a brief review of linear algebra on Euclidean spaces. We shall also recall some simple facts about Euclidean topology. With few exceptions we shall present the results without proofs. General references for linear algebra as needed here are [4], [25], [33] and [53]. An excellent classical reference for the elementary results on analysis we use is [50].

A.1   Vector Spaces

We start by describing the type of vector spaces we shall consider in this book.

A.1.1   Ordered Tuples

Let n be a positive integer. We denote by $\mathbb{R}^n$ the set of all ordered n-tuples

$$\mathbf{x} = (x_1, x_2 \ldots, x_n),$$

where $x_1, \ldots, x_n$ are real numbers, which we call the *coordinates* of $\mathbf{x}$. The elements of $\mathbb{R}^n$ are called *vectors*. For notational convenience we shall denote vectors by bold face letters such as $\mathbf{x}, \mathbf{y}$, and $\mathbf{z}$. In the context of vectors, real numbers are sometimes called *scalars* and are usually denoted by Greek letters such as λ, μ, ν, etc.

The *sum* of two vectors $\mathbf{x}, \mathbf{y} \in \mathbb{R}^n$ is defined by setting

$$\mathbf{x} + \mathbf{y} = (x_1 + y_1, \ldots, x_n + y_n).$$

Thus, the sum of two vectors in $\mathbb{R}^k$ is again a vector in $\mathbb{R}^n$ whose coordinates are obtained simply by coordinate-wise addition of the original vectors. The *multiplication* of a vector $\mathbf{x} \in \mathbb{R}^n$ by any scalar λ is defined by setting

$$\lambda\mathbf{x} = (\lambda x_1, \ldots, \lambda x_n) \ .$$

Thus, multiplication of a vector in $\mathbb{R}^n$ by a scalar again gives a vector in $\mathbb{R}^n$ whose coordinates are obtained by multiplying each coordinate of the original vector by that scalar.

Note that we have not yet defined any multiplication of two vectors.

A.1.2   Vector Space Axioms

The set $V = \mathbb{R}^n$ together with the two operations defined above is an example of a so-called *(real) vector space*. This means that the following eight axioms are satisfied:

a) For every $\mathbf{x}, \mathbf{y} \in V$ we have $\mathbf{x} + \mathbf{y} = \mathbf{y} + \mathbf{x}$ *(commutative law)*.

b) For every $\mathbf{x}, \mathbf{y}, \mathbf{z} \in V$ we have $(\mathbf{x} + \mathbf{y}) + \mathbf{z} = \mathbf{y} + (\mathbf{x} + \mathbf{z})$ *(associative law)*.

c) There exists an element $\mathbf{0} \in V$ such that $\mathbf{x} + \mathbf{0} = \mathbf{x}$ for every $\mathbf{x} \in V$ *(existence of a zero element)*.

d) For every $\mathbf{x} \in V$ there exists an element $\hat{\mathbf{x}} \in V$ such that $\mathbf{x} + \hat{\mathbf{x}} = \mathbf{0}$. For every $\mathbf{x}, \mathbf{y} \in V$, we write $-\mathbf{x}$ and $\mathbf{y} - \mathbf{x}$ instead of $\hat{\mathbf{x}}$ and $\mathbf{y} + \hat{\mathbf{x}}$, respectively *(existence of an inverse)*.

e) For $\lambda, \mu \in \mathbb{R}$ and $\mathbf{x} \in V$ we have $(\lambda + \mu)\mathbf{x} = \lambda\mathbf{x} + \mu\mathbf{x}$ *(distributive law for addition of scalars)*.

f) For $\lambda, \mu \in \mathbb{R}$ and $\mathbf{x} \in V$ we have $(\lambda\mu)\mathbf{x} = \lambda(\mu\mathbf{x})$ *(distributive law for multiplication of scalars)*.

g) For $\lambda \in \mathbb{R}$ and $\mathbf{x}, \mathbf{y} \in V$ we have $\lambda(\mathbf{x} + \mathbf{y}) = \lambda\mathbf{x} + \lambda\mathbf{y}$ *(distributive law for addition of vectors)*.

h) For $\lambda = 1$ we have $\lambda\mathbf{x} = \mathbf{x}$ for every $\mathbf{x} \in V$ *(normalization of scalar multiplication)*.

Any set V on which an "addition" and a "multiplication by scalars" satisfying the above axioms is defined is called a *real vector space*. In case of $\mathbb{R}^n$ the zero element is given by

$$\mathbf{0} = (0, \ldots, 0) \ ,$$

while the inverse of a vector $\mathbf{x} \in \mathbb{R}^n$ is given by

$$-\mathbf{x} = (-x_1, \ldots, -x_n) \ .$$

A.1.3 Linear Subspaces

Let $\mathbf{X}$ be a subset of $\mathbb{R}^n$. Then, $\mathbf{X}$ is a *(linear) subspace* of $\mathbb{R}^n$ if it is itself a vector space, i.e. if the inherited operations satisfy the vector space axioms . For this to be true it is necessary and sufficient that $\mathbf{X}$ be *closed* under the vector space operations, i.e.:

- For $\mathbf{x}, \mathbf{y} \in \mathbf{X}$ we have $\mathbf{x} + \mathbf{y} \in \mathbf{X}$ (*closed under addition*).

- For $\lambda \in \mathbb{R}$ and $\mathbf{x} \in \mathbf{X}$ we have $\lambda \mathbf{x} \in \mathbf{X}$ (*closed under multiplication by scalars*).

For convenience we shall say "Let $\mathbf{X}$ be a vector space" when we actually mean "Let $\mathbf{X}$ be a linear subspace of some $\mathbb{R}^k$".

Linear combinations

Let now $\mathbf{x}_1, \mathbf{x}_2, \ldots, \mathbf{x}_k$ be vectors in $\mathbb{R}^n$ and $\lambda_1, \lambda_2, \ldots, \lambda_k$ be scalars. Then, the vector

$$\sum_{j=1}^{k} \lambda_j \mathbf{x}_j = \lambda_1 \mathbf{x}_1 + \ldots + \lambda_k \mathbf{x}_k$$

is called a *linear combination* of $\mathbf{x}_1, \mathbf{x}_2, \ldots, \mathbf{x}_k$. The scalars $\lambda_1, \lambda_2, \ldots, \lambda_k$ are called the *coefficients* of the linear combination.

The subspace spanned by a subset

Let S be any subset of a vector space $\mathbf{X} \subset \mathbb{R}^n$ and define the *span* of S, *span(S)*, to be the set of all linear combinations of elements of S. It is easy to see that *span(S)* is the smallest linear subspace of $\mathbb{R}^n$ containing S. If $\mathbf{Y}$ is a linear subspace of $\mathbf{X}$ such that $\mathbf{Y} = span(S)$, we shall say that S *spans* $\mathbf{X}$.

A.1.4 Linear Dependence, Bases and Dimension

A subset $\{\mathbf{x}_1, \ldots, \mathbf{x}_k\}$ of a vector space $\mathbf{X} \subset \mathbb{R}^n$ is said to be *(linearly) independent* if the relation $\lambda_1 \mathbf{x}_1 + \ldots + \lambda_k \mathbf{x}_k = 0$ can only be true for $\lambda_1 = \ldots = \lambda_k = 0$. Otherwise, $\{\mathbf{x}_1, \ldots, \mathbf{x}_k\}$ is said to be *(linearly) dependent*.
An independent set of vectors $\{\mathbf{x}_1, \ldots, \mathbf{x}_k\}$ that spans $\mathbf{X}$ is called a *basis* for $\mathbf{X}$.

Theorem A.1 *Let* $\{\mathbf{x}_1, \ldots, \mathbf{x}_k\}$ *be a basis for* $\mathbf{X}$. *Then every vector* $\mathbf{x} \in \mathbf{X}$ *has a unique representation as a linear combination of* $\mathbf{x}_1, \ldots, \mathbf{x}_k$, *i.e. there exist a unique set of scalars* $\lambda_1, \ldots, \lambda_k$, *called the* coordinates *of* $\mathbf{x}$ *(with respect to this basis), such that*

$$\mathbf{x} = \lambda_1 \mathbf{x}_1 + \cdots + \lambda_k \mathbf{x}_k .$$

Standard basis for $\mathbb{R}^n$

Suppose $\mathbf{X} = \mathbb{R}^n$. Denote by $\mathbf{e}_j$ the vector whose j-th coordinate is equal to 1 and all other coordinates are equal to 0, i.e.

$$\mathbf{e}_j = \underbrace{(0,\ldots,0,\overset{\overset{j}{\frown}}{1},0,\ldots,0)}_{k \text{ components}}; j = 1,\ldots,n .$$

The set of vectors $\{\mathbf{e}_1,\ldots,\mathbf{e}_n\}$ is linearly independent. In fact it is a basis for $\mathbb{R}^n$. Indeed, every vector $\mathbf{x} = (x_1,\ldots,x_n)$ can be written as

$$\mathbf{x} = x_1\mathbf{e}_1 + \ldots + x_n\mathbf{e}_n .$$

The set $\{\mathbf{e}_1,\ldots,\mathbf{e}_n\}$ will be called the *standard basis* of $\mathbb{R}^n$. In the case of $n = 1$ the standard basis consists of just one element $\{1\}$.

Basis extension theorem

The following easy result is very useful.

Theorem A.2 *Let $\{\mathbf{x}_1,\ldots,\mathbf{x}_k\}$ be a set of linearly independent vectors in $\mathbf{X}$ and $\{\mathbf{y}_1,\ldots,\mathbf{y}_m\}$ an arbitrary set of vectors. If $span(\{\mathbf{x}_1,\ldots,\mathbf{x}_k,\mathbf{y}_1,\ldots,\mathbf{y}_m\}) = \mathbf{X}$ holds, then by a suitable selection of vectors in $\{\mathbf{y}_1,\ldots,\mathbf{y}_m\}$ we can extend $\{\mathbf{x}_1,\ldots,\mathbf{x}_k\}$ to a basis of $\mathbf{X}$.*

Dimension

In order to introduce the concept of vector space dimension we need the following result.

Theorem A.3 *Let $\{\mathbf{x}_1,\ldots,\mathbf{x}_k\}$ and $\{\mathbf{y}_1,\ldots,\mathbf{y}_m\}$ be bases for $\mathbf{X}$. Then $k = m$.*

Hence, every basis of a vector space $\mathbf{X}$ must have the same cardinality, i.e. the same number of elements. The number of elements any basis of $\mathbf{X}$ must have is called the *dimension* of $\mathbf{X}$ and is denoted by $dim(\mathbf{X})$.
Since we have shown that the set $\{\mathbf{e}_1,\ldots,\mathbf{e}_n\}$ is a basis for $\mathbb{R}^n$ we conclude that the dimension of $\mathbb{R}^n$ is equal to n.

A.1.5   Intersections and Sums of Subspaces

Let $\mathbf{M}_1$ and $\mathbf{M}_2$ be subspaces of $\mathbb{R}^n$. As is easily seen, the intersection $\mathbf{M}_1 \cap \mathbf{M}_2$ of $\mathbf{M}_1$ and $\mathbf{M}_2$ is again a linear subspace.
The *sum* $\mathbf{M}_1 + \mathbf{M}_2$ of $\mathbf{M}_1$ and $\mathbf{M}_2$ is defined by

$$\mathbf{M}_1 + \mathbf{M}_2 \overset{\text{def}}{=} \{\mathbf{x} + \mathbf{y}; \mathbf{x} \in \mathbf{M}_1 \text{ and } \mathbf{y} \in \mathbf{M}_2\} .$$

Lemma A.4 *The following dimension formula holds:*

$$dim(\mathbf{M}_1 \cap \mathbf{M}_2) + dim(\mathbf{M}_1 + \mathbf{M}_2) = dim(\mathbf{M}_1) + dim(\mathbf{M}_2) \,.$$

It follows that if $\mathbf{M}_1 \cap \mathbf{M}_2 = \{\mathbf{0}\}$ we have

$$dim(\mathbf{M}_1 + \mathbf{M}_2) = dim(\mathbf{M}_1) + dim(\mathbf{M}_2) \,.$$

In this case, we say that $\mathbf{M} = \mathbf{M}_1 + \mathbf{M}_2$ is the *direct sum* of $\mathbf{M}_1$ and $\mathbf{M}_2$ and we write

$$\mathbf{M}_1 \oplus \mathbf{M}_2$$

instead of $\mathbf{M}_1 + \mathbf{M}_2$.

A.2 Inner Product and Euclidean Spaces

For two vectors $\mathbf{x}, \mathbf{y} \in \mathbb{R}^n$ we may define their *inner product*, or *scalar product*, by setting

$$\mathbf{x} \cdot \mathbf{y} = (\mathbf{x}|\mathbf{y}) = \sum_{j=1}^{n} x_j y_j = x_1 y_1 + x_2 y_2 + \ldots + x_n y_n \,.$$

Note that the inner product of two vectors is a scalar and *not* a vector. The *norm* of a vector $\mathbf{x} \in \mathbb{R}^n$ is then defined by

$$|\mathbf{x}| = (\mathbf{x}|\mathbf{x})^{1/2} = \left(\sum_{j=1}^{n} x_j^2\right)^{1/2} = (x_1^2 + x_2^2 + \ldots + x_n^2)^{1/2} \,.$$

The set $\mathbb{R}^n$ endowed with the addition of vectors, the multiplication of vectors by scalars and the inner product of two vectors is called *(n-dimensional) Euclidean space*. The following theorem collects some important properties of Euclidean spaces.

Theorem A.5 *Suppose $\mathbf{x}, \mathbf{y}, \mathbf{z} \in \mathbb{R}^n$ and λ is a scalar. Then*

a) $|\mathbf{x}| \geq 0$;

b) $|\mathbf{x}| = 0$ *if and only if* $\mathbf{x} = \mathbf{0}$;

c) $|\lambda \mathbf{x}| = |\lambda||\mathbf{x}|$;

d) $|(\mathbf{x}|\mathbf{y})| \leq |\mathbf{x}|^2 \cdot |\mathbf{y}|^2$ *(Cauchy–Schwarz inequality)*;

e) $|\mathbf{x} + \mathbf{y}| \leq |\mathbf{x}| + |\mathbf{y}|$ *(Triangle inequality)*;

f) $|\mathbf{x} - \mathbf{z}| \leq |\mathbf{x} - \mathbf{y}| + |\mathbf{y} - \mathbf{z}|$.

A.2.1   Angles and Orthogonality

By the Cauchy–Schwarz inequality we have for any $\mathbf{x}, \mathbf{y} \in \mathbb{R}^n$,

$$-1 \leq \frac{(\mathbf{x}|\mathbf{y})}{|\mathbf{x}| \cdot |\mathbf{y}|} \leq 1 \,.$$

Hence, since $\cos : [0, 180] \to [-1, 1]$ is a bijective function, there exists a unique number $\alpha(\mathbf{x}, \mathbf{y}) \in [0, 180]$, called the *angle* formed by $\mathbf{x}$ and $\mathbf{y}$, such that

$$\cos \alpha(\mathbf{x}, \mathbf{y}) = \frac{(\mathbf{x}|\mathbf{y})}{|\mathbf{x}| \cdot |\mathbf{y}|} \,.$$

Hence, $(\mathbf{x}|\mathbf{y}) \geq 0$ is another way of saying that $\mathbf{x}$ and $\mathbf{y}$ form an angle between 0 and 90. By contrast, $(\mathbf{x}|\mathbf{y}) \leq 0$ means that $\mathbf{x}$ and $\mathbf{y}$ form an angle between 90 and 180.

Orthogonality

Two vectors $\mathbf{x}$ and $\mathbf{y}$ are said to be *orthogonal* to each other if they form an angle equal to 90, i.e. if $(\mathbf{x}|\mathbf{y}) = 0$ holds.

Let now $\mathbf{A}$ be a subset of a vector space $\mathbf{X}\mathbb{R}^n$. We define its *orthogonal complement* $\mathbf{A}^\perp$ by

$$\mathbf{A}^\perp \overset{\text{def}}{=} \{\mathbf{x} \in \mathbf{X}; (\mathbf{x}|\mathbf{y}) = 0 \text{ for all } \mathbf{y} \in \mathbf{A}\} \,.$$

It is easy to show that $\mathbf{A}^\perp$ is in fact a linear subspace of $\mathbf{X}$ and that the following result holds .

Lemma A.6 *Let $\mathbf{M}$ be a subspace of $\mathbf{X}$. Then,*

$$\mathbf{X} = \mathbf{M} \oplus \mathbf{M}^\perp \,.$$

It is important to note that we can therefore uniquely decompose each element in $\mathbf{X}$ into its part $\mathbf{x_M}$ in $\mathbf{M}$ and its part in $\mathbf{x_{M^\perp}}$ $\mathbf{M}^\perp$, i.e.

$$\mathbf{x} = \mathbf{x_M} + \mathbf{x_{M^\perp}} \,.$$

A.3   Topology in Euclidean Space

The *distance* $d(\mathbf{x}, \mathbf{y})$ between the vectors $\mathbf{x}$ and $\mathbf{y}$ is defined by

$$d(\mathbf{x}, \mathbf{y}) \overset{\text{def}}{=} |\mathbf{x} - \mathbf{y}| \,.$$

Endowed with this distance any subset $\mathbf{A}$ of $\mathbb{R}^n$ becomes a metric space. Recall the following topological concepts:

- A sequence $(\mathbf{x}_j)$ is said to *converge* in (the relative topology of) $\mathbf{A}$ if there exists a vector $\mathbf{x} \in \mathbf{A}$ such that $\lim_{j \to \infty} d(\mathbf{x}_j, \mathbf{x}) = 0$. In this case $\mathbf{x}$ is said to be the *limit* of the sequence $(\mathbf{x}_j)$.

- $\mathbf{B} \subset \mathbf{A}$ is said to be *open* in (the relative topology of)$\mathbf{A}$ if for any $\mathbf{x}_0 \in \mathbf{B}$ there is an $\epsilon > 0$ such that $\{\mathbf{x} \in \mathbf{A}; d(\mathbf{x}, \mathbf{x}_0) < \epsilon\}$ is contained in $\mathbf{B}$.

- $\mathbf{B} \subset \mathbf{A}$ is *closed* in (the relative topology of) $\mathbf{A}$ if $\mathbf{A} \setminus \mathbf{B}$ is open in $\mathbf{A}$. Equivalently, $\mathbf{B} \subset \mathbf{A}$ is closed in $\mathbf{A}$ if and only if all vectors in $\mathbf{A}$ which are limits of some sequence in $\mathbf{B}$ are contained in $\mathbf{B}$, i.e. if $\mathbf{x}_j \in \mathbf{B}$ and $\mathbf{x}_j \to \mathbf{x} \in \mathbf{A}$ imply that $\mathbf{x} \in \mathbf{B}$.

- $\mathbf{A} \subset \mathbb{R}^n$ is said to be *compact* if it is closed in $\mathbb{R}^n$ and every sequence $(\mathbf{x}_j)$ in $\mathbf{A}$ has a convergent subsequence. Equivalently, $\mathbf{A}$ is compact if and only if it is closed in $\mathbb{R}^n$ and bounded.

Continuous functions

Let $\mathbf{A}$ be a subset of $\mathbb{R}^n$ and $f : \mathbf{A} \to \mathbb{R}^m$ a function. Recall that f is called continuous if for each sequence $(\mathbf{x}_j)$ in $\mathbf{A}$ converging to a vector $\mathbf{x} \in \mathbf{A}$ we have that the sequence $(f(\mathbf{x}_j))$ converges to $f(\mathbf{x})$ in $\mathbb{R}^m$.

Proposition A.7 *A continuous function $f : \mathbf{A} \to \mathbb{R}^m$ defined on a compact subset $\mathbf{A}$ of $\mathbb{R}^n$ has a minimum and a maximum, i.e. there exist $\mathbf{x}_{min}$ and $\mathbf{x}_{max}$ in $\mathbf{A}$ such that*

$$f(\mathbf{x}_{min}) \leq f(\mathbf{x}) \leq f(\mathbf{x}_{max})$$

holds for all $\mathbf{x} \in \mathbf{A}$.

Combining two continuous functions by means of a variety of elementary operations preserves continuity, as stated next in the following elementary result.

Proposition A.8 *Let $f, g : \mathbf{A} \to \mathbb{R}^m$ be continuous functions. Then, $f + g$ is also continuous.*
If we assume that f, g are real valued (i.e. m=1), then $f \cdot g$ and $\max\{f, g\}$ are continuous too. The same is true for f/g, if $g(s) \neq 0$ for all $s \in \mathbf{A}$.

Monotonous and convex functions

Let $g : I \to \mathbb{R}$ be a function defined on some interval I in $\mathbb{R}$. Then, g is *decreasing* if for $x \leq y$ we have

$$g(x) \geq g(y) .$$

If above we have strict inequality for all $x < y$, we say that g is *strictly* increasing. We say g is *(strictly) increasing* if $-g$ is (strictly) decreasing. If g is either increasing or decreasing we say that g is a *monotonous* function.

The function g is convex if for $x, y \in I$ and $\lambda \in [0, 1]$ we have

$$g(\lambda x + (1 - \lambda)y) \leq \lambda g(x) + (1 - \lambda)g(y) \, .$$

We say g is *concave* if $-g$ is convex.

Taking the maximum of two decreasing or convex functions preserves the property as stated in the following easy to prove result.

Proposition A.9 *Let $f, g : I \to \mathbb{R}$ be two functions and define*

$$h(x) \stackrel{def}{=} \max\{g(x), f(x)\} \, .$$

Then, if f and g are both (strictly) decreasing, then so is h. If f and g are both convex, then so is h.

A similar result holds for increasing or concave functions if we takes the minimum instead of the maximum.

A.4   Linear Operators

Let $\mathbf{X}$ and $\mathbf{Y}$ be two vector spaces. We denote by n and by m the dimensions of $\mathbf{X}$ and $\mathbf{Y}$, respectively. A function $A : \mathbf{X} \to \mathbf{Y}$ is said to be a *linear operator* [1] from $\mathbf{X}$ into $\mathbf{Y}$ if it satisfies the two properties:

a) $A(\mathbf{x} + \mathbf{y}) = A\mathbf{x} + A\mathbf{y}$ (*additivity*),

b) $A(\lambda \mathbf{x}) = \lambda A\mathbf{x}$ (*homogeneity*)

for all vectors $\mathbf{x}, \mathbf{y} \in \mathbf{X}$ and all scalars λ. Here we have adopted the convention of writing $A\mathbf{x}$ instead of $A(\mathbf{x})$ when no confusion seems likely. When invoking either of the two characteristic properties of linear operators we shall use the phrase "by linearity".

Thus, a function $A : \mathbf{X} \to \mathbf{Y}$ is a linear operator if it is compatible with the vector space operations, i.e. it makes no difference whether we first add two vectors $\mathbf{x}$ and $\mathbf{y}$ and then apply A to the sum $\mathbf{x} + \mathbf{y}$, or if we first apply A to $\mathbf{x}$ and $\mathbf{y}$ separately and then add the results; and analogously for multiplication with scalars. One also says that linear operators *preserve* the vector space structure.

A few simple remarks are in order. Let $A : \mathbf{X} \to \mathbf{Y}$ be a linear operator. Then:

a) $A\mathbf{0} = \mathbf{0}$, where, naturally, the first $\mathbf{0}$ is the zero element in $\mathbf{X}$ and the second the one in $\mathbf{Y}$.

---

[1]The terms linear *map, mapping, transformation, function, homomorphism* are all used interchangeably throughout the literature.

b) A is completely determined by its values on any basis of $\mathbf{X}$. Indeed, let $\{\mathbf{x}_1, \ldots, \mathbf{x}_n\}$ be a basis of $\mathbf{X}$. Then any $\mathbf{x} \in \mathbf{X}$ can be written as $\mathbf{x} = \lambda_1 \mathbf{x}_1 + \ldots + \lambda_n \mathbf{x}_n$. By linearity we have

$$A\mathbf{x} = \lambda_1 A\mathbf{x}_1 + \ldots + \lambda_n A\mathbf{x}_n .$$

Hence, this formula allows us to compute $A\mathbf{x}$ from our knowledge of the coordinates of $\mathbf{x}$ with respect to the basis and the values $A\mathbf{x}_1, \ldots, A\mathbf{x}_n$.

A.4.1 Continuity of Linear Operators

In finite dimensional vector spaces all linear operators are continuous as is recorded in the next proposition.

Proposition A.10 *Let $A : \mathbf{X} \to \mathbf{Y}$ be a linear operator where $\mathbf{X}$ and $\mathbf{Y}$ are given vector spaces. Then, A is continuous.*

A.4.2 Orthogonal Projections

Let $\mathbf{M}$ be a subspace of $\mathbf{X}$. We considered its orthogonal complement $\mathbf{M}^\perp$ and noted that each vector $\mathbf{x}$ in $\mathbf{X}$ could be written in a unique manner as the sum of a vector $\mathbf{x}_\mathbf{M}$ in $\mathbf{M}$ and a vector $\mathbf{x}_{\mathbf{M}^\perp}$ in $\mathbf{M}^\perp$. Define the function $P_\mathbf{M} : \mathbf{X} \to \mathbf{M}$ by setting

$$P_\mathbf{M}(\mathbf{x}) \overset{\text{def}}{=} \mathbf{x}_\mathbf{M} .$$

Proposition A.11 *$P_\mathbf{M} : \mathbf{X} \to \mathbf{M}$ is a linear operator satisfying $P_\mathbf{M}(\mathbf{x}) = \mathbf{x}$ for all $\mathbf{x} \in \mathbf{M}$ and $P_\mathbf{M}(\mathbf{x}) = 0$ for all $\mathbf{x} \in \mathbf{M}^\perp$. Moreover, it is the only linear operator satisfying these two conditions. $P_\mathbf{M}$ is called the* orthogonal projection *onto* $\mathbf{M}$.

A.4.3 Operations on Linear Operators

We denote by $\mathcal{L}(\mathbf{X}, \mathbf{Y})$ the set of all linear operators from $\mathbf{X}$ to $\mathbf{Y}$. If $\mathbf{X} = \mathbf{Y}$ we shall simply write $\mathcal{L}(\mathbf{X})$. For any $A, B \in \mathcal{L}(\mathbf{X}, \mathbf{Y})$ define their sum by

$$(A + B)\mathbf{x} = A\mathbf{x} + B\mathbf{x}, \qquad \mathbf{x} \in \mathbf{X} .$$

Furthermore, we define the multiplication of A by any scalar λ by

$$(\lambda A)\mathbf{x} = \lambda(A\mathbf{x}), \qquad \mathbf{x} \in \mathbf{X} .$$

It is easy to verify that $A + B$ and λA are both linear operators in $\mathcal{L}(\mathbf{X}, \mathbf{Y})$.

Remark A.12 *The set $\mathcal{L}(\mathbf{X}, \mathbf{Y})$ endowed with the above defined addition and multiplication with scalars satisfies the vector space axioms and is therefore a vector space.*

If $\mathbf{Z}$ is a further vector space we may define for any $A \in \mathcal{L}(\mathbf{X}, \mathbf{Y})$ and $B \in \mathcal{L}(\mathbf{Y}, \mathbf{Z})$ their *composition* by

$$(BA)\mathbf{x} = B(A\mathbf{x}), \qquad \mathbf{x} \in \mathbf{X}.$$

Assuming that $\mathbf{X} = \mathbf{Y} = \mathbf{Z}$, both AB and BA belong to $\mathcal{L}(\mathbf{X})$. Note, however, that in general $AB \neq BA$!

A.5   Linear Equations

Many problems in economics and other disciplines which can be cast into the language of linear algebra are formulated as linear equations. More precisely, we are given a pair of vector spaces $\mathbf{X}$ and $\mathbf{Y}$, a linear operator $A : \mathbf{X} \to \mathbf{Y}$ and a vector $\mathbf{b} \in \mathbf{Y}$ and are asked to find the set of all $\mathbf{x} \in \mathbf{X}$ satisfying the *linear equation*

$$A\mathbf{x} = \mathbf{b}. \tag{A.1}$$

The set of solutions of the above equation will be denoted by $\mathcal{S}(A, \mathbf{b})$, i.e.

$$\mathcal{S}(A, \mathbf{b}) = \{\mathbf{x} \in \mathbf{X}; A\mathbf{x} = \mathbf{b}\}.$$

The range of a linear operator

We first look at the set of vectors $\mathbf{b}$ for which the above linear equation admits a solution. For any $\mathbf{x} \in \mathbf{X}$, the vector $A\mathbf{x}$ is called the *image* of $\mathbf{x}$ under A. It is clear that (A.1 admits a solution if and only if $\mathbf{b}$ is the image of some $\mathbf{x} \in \mathbf{X}$. If $\mathbf{S}$ is any subset of $\mathbf{X}$ we shall write $A(\mathbf{S})$ for the *image of* $\mathbf{S}$ *under* A, i.e. for the set consisting of all images of vectors in $\mathbf{S}$. The image of $\mathbf{X}$ under A is called the *range* of A and will be denoted by $R(A)$.

Proposition A.13 *For any linear operator* $A : \mathbf{X} \to \mathbf{Y}$, *its range* $R(A)$ *is a linear subspace of* $\mathbf{Y}$. *If* $\mathbf{B} = \{\mathbf{x}_1, \ldots, \mathbf{x}_n\}$ *is a basis for* $\mathbf{X}$, *then* $R(A) = span(A(\mathbf{B}))$.

The kernel of a linear operator

The solution set of the equation

$$A\mathbf{x} = \mathbf{0}$$

is called the *kernel* or *null space* of A and will be denoted by $N(A)$, i.e. $N(A) = \mathcal{S}(A, \mathbf{0})$. The kernel of A is thus the set of all vectors whose image is $\mathbf{0}$. It is easy to verify that if $\mathbf{x}, \mathbf{y} \in N(A)$ and λ is a scalar, then $\mathbf{x} + \mathbf{y}$ and $\lambda \mathbf{x}$ both again belong to $N(A)$. Therefore, we have

Proposition A.14 $N(A)$ *is a linear subspace of* $\mathbf{X}$.

Structure of the solution set

The above proposition describes the structure of the solution set of (A.1) if $\mathbf{b} = 0$. More generally, if $\mathbf{x}$ is a *particular solution*, i.e. a particular element of $\mathcal{S}(A, \mathbf{b})$, and $\mathbf{y} \in N(A) = \mathcal{S}(A, 0)$, then we have that $\mathbf{x} + \mathbf{y}$ belongs to $\mathcal{S}(A, \mathbf{b})$. In fact, every further solution of (A.1) can be obtained in this manner:

Proposition A.15 *If $\hat{\mathbf{x}}$ is a particular solution of (A.1), then*

$$\mathcal{S}(A, \mathbf{b}) = \hat{\mathbf{x}} + N(A) .$$

This means that every other solution $\mathbf{x}$ of (A.1) can be written as $\mathbf{x} = \hat{\mathbf{x}} + \mathbf{y}$ with $\mathbf{y} \in N(A)$.

The above result offers an important insight on the solution set of (A.1). In order to solve (A.1) we need only determine a particular solution and the set $ker(A)$.

The dimension formula for linear operators

We have the following relationship between the dimensions of the kernel and the range of a linear operator.

Proposition A.16 *For any linear operator $A : \mathbf{X} \to \mathbf{Y}$ we have:*

$$dim(X) = dim(N(A)) + dim(R(A)) .$$

A.5.1 Injectivity, Surjectivity and Bijectivity

A linear operator $A : \mathbf{X} \to \mathbf{Y}$ is said to be *injective* if $\mathbf{x} \neq \mathbf{y}$ implies $A\mathbf{x} \neq A\mathbf{y}$, i.e. if distinct vectors have distinct images under A.
The following result establishes a relation between the solvability of the linear equation $A\mathbf{x} = \mathbf{0}$ and injectivity.

Proposition A.17 *The linear operator $A : \mathbf{X} \to \mathbf{Y}$ is injective if and only if its kernel is* trivial, *i.e. if $N(A) = \{0\}$.*
Moreover, if A is injective it is necessary that $dim(\mathbf{X}) \leq dim(\mathbf{Y})$.

A linear operator $A : \mathbf{X} \to \mathbf{Y}$ is *surjective* if for each $\mathbf{b} \in \mathbf{Y}$ the linear equation $A\mathbf{x} = \mathbf{b}$ is solvable, i.e. if there exists $\mathbf{x} \in \mathbf{X}$ such that $A\mathbf{x} = \mathbf{b}$. Another way of putting this is by saying that A is surjective if and only if the range of A is the whole of $\mathbf{Y}$.

Proposition A.18 *For A to be surjective it is necessary that $dim(\mathbf{X}) \geq dim(\mathbf{Y})$.*

We call A *bijective* if it is both injective and surjective. This means that for each $\mathbf{b} \in \mathbf{Y}$ the linear equation $A\mathbf{x} = \mathbf{b}$ is uniquely solvable.

Proposition A.19 *If A is bijective it is necessary that $dim(\mathbf{X}) = dim(\mathbf{Y})$. More-over, if $dim(\mathbf{X}) = dim(\mathbf{Y})$ then A is injective if and only if it is surjective.*

It follows that when both spaces $\mathbf{X}$ and $\mathbf{Y}$ are of the same dimension, injectivity, surjectivity and bijectivity are all equivalent.

Inverse operator

In the case of an injective operator we may define the *inverse operator* A^{-1}: If $\mathbf{v} \in \mathbf{Y}$ we set $A^{-1}\mathbf{v} = \mathbf{x}$ where $\mathbf{x} \in \mathbf{X}$ is such that $A\mathbf{x} = \mathbf{v}$. Such an $\mathbf{x}$ exists by the surjectivity of A and is unique by the injectivity. The inverse operator A^{-1} can be also characterized as the only operator satisfying

$$A^{-1}A\mathbf{x} = \mathbf{x} \qquad \text{and} \qquad AA^{-1}\mathbf{v} = \mathbf{v}$$

for all $\mathbf{x} \in \mathbf{X}$ and $\mathbf{v} \in \mathbf{Y}$.

Appendix B

Proof of the Theorem of de Moivre–Laplace

In this Appendix we prove a special version of the Central Limit Theorem, known as the Theorem of de Moivre–Laplace. This theorem was already stated as Theorem 13.23 and captures and generalizes the observations made on averages of averages in the coin tossing experiment considered at the beginning of Chapter 13 in a rigorous mathematical language. The proof of the theorem is quite accessible since only basic tools of calculus need to be applied.

Theorem B.1 (de Moivre–Laplace) *Consider a sequence $p_n \in (0,1)$ such that $p_n \to p \in (0,1)$ and let $(S_n)_{n \in \mathbb{N}}$ denote a sequence of binomial random variables with parameters n and p_n. Then the sequence of re-scaled random variables*

$$S_n^* := \frac{S_n - E[S_n]}{\sqrt{Var(S_n)}} = \frac{S_n - n\,p_n}{\sqrt{n\,p_n\,(1 - p_n)}}$$

converges weakly to the standard normal random variable, i.e.

$$\lim_{n \to \infty} F_{S_n^*}(s) = F_{\mathcal{N}}(s)$$

for each $s \in \mathbb{R}$.

Before proving this theorem we will prove two weaker versions in Propositions B.4 and B.5.

B.1   Preliminary results

We will use the following notation for the probability densities of the standard normal and the binomial distributions, respectively:

$$\varphi(x) := \frac{1}{\sqrt{2\pi}} e^{-\frac{x^2}{2}} \ , x \in \mathbb{R}$$

and

$$B_{n,p}(k) := \left\{ \begin{array}{ll} \binom{n}{k} p^k (1-p)^{n-k} \ , & k \in \mathbb{Z} \ , k \geq 0 \ , \\ 0, & k \in \mathbb{Z} \ , k < 0 \ . \end{array} \right.$$

Furthermore, we will apply the following version of Taylor's Theorem.

Lemma B.2 *Consider a real-valued function f on an interval (a,b). Assume that the third derivative f''' of f exists, is continuous on (a,b), and satisfies the inequality*

$$f'''(t) \leq M$$

for $t \in (a,b)$. Then f can be written as

$$f(t) = f(t_0) + f'(t_0)(t - t_0) + \frac{1}{2} f''(t_0) (t - t_0)^2 + R(t - t_0)$$

with

$$|R(t - t_0)| \leq \frac{M}{3!} |t - t_0|^3 \ .$$

Some technical observations are collected in the following lemma.

Lemma B.3 *Consider a sequence $(p_n) \in (0,1)$ with $p_n \to p \in (0,1)$ and sequences (α_n) and (β_n) of integers with $\alpha_n \leq \beta_n$ such that*

$$\frac{1}{\sqrt{n}} \left(\frac{\alpha_n - np_n}{\sigma_n} \right)^3 \to 0 \ , \ \text{as } n \to \infty \ ,$$

and

$$\frac{1}{\sqrt{n}} \left(\frac{\beta_n - np_n}{\sigma_n} \right)^3 \to 0 \ , \ \text{as } n \to \infty \ ,$$

where $\sigma_n := \sqrt{np_n q_n}$ and $q_n := 1 - p_n$. Then,

$$\lim_{n \to \infty} \frac{B_{n,p_n}(k_n)}{\frac{1}{\sigma_n} \varphi\left(\frac{k_n - np_n}{\sigma_n} \right)} = 1$$

uniformly for sequences (k_n) with $\alpha_n \leq k_n \leq \beta_n$.

Proof We have to prove that for any $\varepsilon > 0$ we can find $N(\varepsilon) \geq 0$ such that

$$1 - \varepsilon \leq \frac{B_{n,p_n}(k_n)}{\frac{1}{\sigma_n}\varphi\left(\frac{k_n - np_n}{\sigma_n}\right)} \leq 1 + \varepsilon$$

for $n \geq N(\varepsilon)$ where k_n is any sequence with $\alpha_n \leq k_n \leq \beta_n$. Before beginning with the main elements of the proof, we observe that the assumption entails that

$$\lim_{n \to \infty} \frac{1}{\sqrt{n}} \frac{(k_n - np_n)^3}{(np_n q_n)^{\frac{3}{2}}} = \frac{1}{(p_n q_n)^{\frac{3}{2}}} \lim_{n \to \infty} n \left(\frac{k_n}{n} - p_n\right)^3 = 0$$

where the convergence is uniform for sequences $\alpha_n \leq k_n \leq \beta_n$. As a consequence $\lim_{n \to \infty} (\frac{k_n}{n} - p_n)^3 = 0$ and by the convergence of p_n we also find that

$$\lim_{n \to \infty} \frac{k_n}{n} = p \,, \quad \lim_{n \to \infty} \frac{k_n}{np_n} = 1 \,, \quad \lim_{n \to \infty} \frac{n - k_n}{n(1 - p_n)} = 1 \,. \tag{B.1}$$

We subdivide the proof into several steps:

Step I: Applying the Stirling approximation

$$\lim_{n \to \infty} \frac{n!}{\sqrt{2\pi n} \left(\frac{n}{e}\right)^n} = 1$$

to the binomial coefficient in $B_{n,p_n}(k_n)$ we obtain its asymptotic behavior for n tending to infinity

$$\binom{n}{k_n} = \frac{n!}{k_n!(n - k_n)!} \sim \frac{1}{\sqrt{2\pi}} \sqrt{\frac{n}{k_n(n - k_n)}} \frac{n^{k_n} n^{n - k_n}}{k_n^{k_n} (n - k_n)^{n - k_n}} \,.$$

Consequently,

$$\lim_{n \to \infty} \frac{B_{n,p_n}(k_n)}{\frac{1}{\sqrt{2\pi}} \sqrt{\frac{n}{k_n(n - k_n)}} \left(\frac{np_n}{k_n}\right)^{k_n} \left(\frac{n(1 - p_n)}{n - k_n}\right)^{n - k_n}} = 1 \,. \tag{B.2}$$

Step II: By $(B.1)$ we can easily determine the asymptotic behavior of one of the factors in the denominator of $(B.2)$ as

$$\lim_{n \to \infty} \frac{\sqrt{\frac{n}{k_n(n - k_n)}}}{\frac{1}{\sigma_n}} = 1 \,.$$

Step III: We study the asymptotic behavior of the other factor in the denominator of $(B.2)$

$$h(k_n, n) := \left(\frac{np_n}{k_n}\right)^{k_n} \left(\frac{n(1 - p_n)}{n - k_n}\right)^{n - k_n} \,.$$

We verify that

$$\lim_{n\to\infty} \frac{h(k_n, n)}{e^{-\frac{1}{2}(\frac{k_n - np_n}{\sigma_n})^2}} = 1 .$$

In fact, setting $t_n := \frac{k_n}{n}$ and taking the natural logarithm we introduce the function g as

$$g(t_n) := -\frac{1}{n} \ln (h(k_n, n)) = \left[t_n \ln \left(\frac{t_n}{p_n} \right) + (1 - t_n) \ln \left(\frac{1 - t_n}{q_n} \right) \right] .$$

Expanding $g(t)$ into its Taylor series around p_n (verify that $g(p_n) = g'(p_n) = 0$ and $g''(p_n) = \frac{1}{p_n q_n}$) we see that

$$g(t) = \frac{1}{2p_n q_n}(t - p_n)^2 + R(t - p_n) ,$$

where the function R satisfies

$$|R(t - p_n)| \le c|t - p_n|^3$$

for some positive constant c and each t in a neighborhood of p_n. Using the assumption $n(t_n - p_n)^3 \to 0$ as $n \to \infty$ we conclude that

$$n R(t_n - p_n) \to 0 \quad \text{as } n \to \infty$$

where again the convergence is uniform for sequences $k_n = n t_n$ with $\alpha_n \le k_n \le \beta_n$. Hence,

$$|n g(t_n) - \frac{n}{2p_n q_n} (t_n - p_n)^2| \to 0 , \quad \text{as } n \to \infty$$

and since

$$\frac{n}{2p_n q_n}(t_n - p_n)^2 = \frac{1}{2} \left(\frac{k_n - np_n}{\sigma_n} \right)^2$$

we obtain the desired relation

$$\lim_{n\to\infty} \frac{h(k_n, n)}{e^{-\frac{1}{2}(\frac{k_n - np_n}{\sigma_n})^2}} = 1 .$$

Finally, as a consequence of the above steps I, II, and III we conclude that

$$\lim_{n\to\infty} \frac{B_{n,p_n}(k_n)}{\frac{1}{\sigma_n \sqrt{2\pi}}e^{-\frac{1}{2}(\frac{k_n - np_n}{\sigma_n})^2}} = 1 ,$$

i.e.

$$\lim_{n\to\infty} \frac{B_{n,p_n}(k_n)}{\frac{1}{\sigma_n}\varphi(\frac{k_n - np_n}{\sigma_n})} = 1 ,$$

uniformly for sequences k_n with $\alpha_n \le k_n \le \beta_n$. $\qquad\square$

Proposition B.4 *For any choice of $a, b \in \mathbb{R}$ with $a < b$ we have*

$$\lim_{n \to \infty} P\left(a \le S_n^* \le b\right) = F_{\mathcal{N}}(b) - F_{\mathcal{N}}(a) = \frac{1}{\sqrt{2\pi}} \int_a^b e^{\frac{-x^2}{2}} \, dx \; .$$

Proof For $n \in \mathbb{N}$ set

$$\alpha_n := \lceil a\sigma_n + np_n \rceil \quad \text{and} \quad \beta_n := \lfloor b\sigma_n + np_n \rfloor \; .$$

It is trivial to verify that

$$P\left(a \le S_n^* \le b\right) = P\left(\alpha_n \le S_n \le \beta_n\right)$$

and by definition

$$P\left(\alpha_n \le S_n \le \beta_n\right) = \sum_{k=\alpha_n}^{\beta_n} B_{n,p_n}(k) \; . \tag{B.3}$$

At this point we apply Lemma B.3, hence, we must verify the assumption

$$\lim_{n \to \infty} \frac{1}{\sqrt{n}} \left(\frac{k_n - np_n}{\sqrt{np_nq_n}}\right)^3 = 0$$

or equivalently

$$\lim_{n \to \infty} n \left(\frac{k_n}{n} - p_n\right)^3 = 0$$

for any sequence (k_n) satisfying

$$\alpha_n \le k_n \le \beta_n \, , \; n \in \mathbb{N} \; .$$

Indeed, the condition is satisfied as a consequence of the following elementary observations

$$\lim_{n \to \infty} n \left(\frac{\alpha_n}{n} - p_n\right)^3 = \lim_{n \to \infty} n \left(\frac{\lceil a\sigma_n + np_n \rceil}{n} - p_n\right)^3 = \lim_{n \to \infty} \frac{1}{n^2} \left(\lceil a\sigma_n + np_n \rceil - np_n\right)^3$$

$$\le \lim_{n \to \infty} \frac{1}{n^2} \left(a\sqrt{np_nq_n} + np_n + 1 - np_n\right)^3 = \lim_{n \to \infty} a \left(p_nq_n\right)^{\frac{3}{2}} \frac{n^{\frac{3}{2}}}{n^2} = 0.$$

Analogously one can verify that

$$\lim_{n \to \infty} n \left(\frac{\beta_n}{n} - p_n\right)^3 = 0 \; .$$

Hence, by Lemma B.3, for any $\varepsilon > 0$ we can find $N(\varepsilon)$ such that

$$(1 - \varepsilon)\frac{1}{\sigma_n}\varphi(\frac{k - np_n}{\sigma_n}) \leq B_{n,p_n}(k) \leq (1 + \varepsilon)\frac{1}{\sigma_n}\varphi(\frac{k - np_n}{\sigma_n})$$

for all k with $\alpha_n \leq k \leq \beta_n$ and all $n \geq N(\varepsilon)$. Hence, for $n \geq N(\varepsilon)$ we obtain that

$$(1 - \varepsilon)\sum_{k=\alpha_n}^{\beta_n}\frac{1}{\sigma_n}\varphi(\frac{k - np_n}{\sigma_n}) \leq \sum_{k=\alpha_n}^{\beta_n}B_{n,p_n}(k) \leq (1 + \varepsilon)\sum_{k=\alpha_n}^{\beta_n}\frac{1}{\sigma_n}\varphi(\frac{k - np_n}{\sigma_n})$$

i.e.

$$\lim_{n \to \infty}\frac{\sum_{k=\alpha_n}^{\beta_n}B_{n,p_n}(k)}{\sum_{k=\alpha_n}^{\beta_n}\frac{1}{\sigma_n}\varphi(\frac{k - np_n}{\sigma_n})} = 1. \tag{B.4}$$

Next observe that

$$\sum_{k=\alpha_n}^{\beta_n}\frac{1}{\sigma_n}\varphi(\frac{k - np_n}{\sigma_n}) = \sum_{l=0}^{N(n)}\frac{1}{\sigma_n}\varphi\left(\frac{\alpha_n - np_n}{\sigma_n} + \frac{l}{\sigma_n}\right), \tag{B.5}$$

where $N(n) := \beta_n - \alpha_n \to \infty$ as $n \to \infty$. The right-hand side in $(B.5)$ is a Riemann sum approximating the integral

$$\int_a^b \varphi(x)dx,$$

i.e.

$$\lim_{n \to \infty}\sum_{l=0}^{N(n)}\frac{1}{\sigma_n}\varphi\left(\frac{\alpha_n - np_n}{\sigma_n} + \frac{l}{\sigma_n}\right) = \int_a^b \varphi(x)dx. \tag{B.6}$$

Here we have made use of the simple observations

$$\lim_{n \to \infty}\frac{\alpha_n - np_n}{\sigma_n} = \lim_{n \to \infty}\frac{\lceil a\sigma_n + np_n \rceil - np_n}{\sigma_n} = a$$

and

$$\lim_{n \to \infty}\frac{\beta_n - np_n}{\sigma_n} = \lim_{n \to \infty}\frac{\lfloor b\sigma_n + np_n \rfloor - np_n}{\sigma_n} = b.$$

Finally from $(B.3)$, $(B.4)$, $(B.5)$, and $(B.6)$ it is easily inferred that

$$\lim_{n \to \infty}P\left(a \leq S_n^* \leq b\right) = \int_a^b \varphi(x)\,dx$$

for any $a, b \in \mathbb{R}$, with $a < b$. $\square$

In the next proposition we show that the convergence of $P(a \leq S_n^* \leq b)$ is uniform with respect to the choice of a, b.

Proposition B.5

$$\lim_{n \to \infty} \sup_{a < b} \left| P(a \leq S_n^* \leq b) - \int_a^b \varphi(x) dx \right| = 0 \,.$$

Proof This is equivalent to showing that to arbitrary $\varepsilon > 0$ there exists $N(\varepsilon) \in \mathbb{N}$ such that for all $a, b \in \mathbb{R}$ with $a < b$,

$$\int_a^b \varphi(x) dx - \varepsilon \leq P(a \leq S_n^* \leq b) \leq \int_a^b \varphi(x) dx + \varepsilon \tag{B.7}$$

for $n \geq N(\varepsilon)$. Here, the crucial point is that $N(\varepsilon)$ must only depend on ε and not on the particular choice of a, b.

Fix an arbitrary $\varepsilon > 0$. Since $\int_{-\infty}^{\infty} \varphi(x) dx = 1$ it is possible to choose $t_1, \ldots, t_m \in \mathbb{R}$ with

$$t_j < t_{j+1}, \quad j \in \{1, \ldots, m-1\}$$

such that

$$\int_{t_j}^{t_{j+1}} \varphi(x) dx < \varepsilon, \, j \in \{1, \ldots, m-1\}, \tag{B.8}$$

and

$$\int_{-\infty}^{t_1} \varphi(x) dx + \int_{t_m}^{\infty} \varphi(x) dx < \varepsilon. \tag{B.9}$$

As a consequence of Proposition B.4 we can choose $N(\varepsilon) \in \mathbb{N}$ such that for $n \geq N(\varepsilon)$ and $(a, b) \in [t_1, t_m]$ with $a < b$ we have

$$\left| P(a \leq S_n^* \leq b) - \int_a^b \varphi(x) \, dx \right| < \varepsilon. \tag{B.10}$$

For arbitrary $a, b \in \mathbb{R}$ we can write the interval $[a, b]$ as the disjoint union

$$[a, b] = I_1 \cup I_2 \cup I_3$$

where the possibly empty sets I_i are defined as

$$I_1 := [a, b] \cap (-\infty, t_1], \, I_2 := [a, b] \cap (t_1, t_m], \, I_3 := [a, b] \cap (t_m, \infty).$$

Since

$$P(a \leq S_n^* \leq b) = P(S_n^* \in [a, b]) = P(S_n^* \in I_1) + P(S_n^* \in I_2) + P(S_n^* \in I_3) \,,$$

we will derive estimates for $P(S_n^* \in I_i)$, $i \in \{1, 2, 3\}$ of the form

$$\int_{a_i'}^{b_i'} \varphi(x)\, dx + \varepsilon \geq P(S_n^* \in I_i) \geq \int_{a_i'}^{b_i'} \varphi(x)\, dx - \varepsilon$$

where a_i' and b_i' are the boundary points of the interval I_i .
We first show the inequality

$$P(a \leq S_n^* \leq b) \geq \int_{a}^{b} \varphi(x)dx - \varepsilon$$

for n sufficiently large and any a, b with $a < b$:

Since $I_1 \subset [a_1', b_1'] \subset (-\infty, t_1]$ and $I_3 \subset [a_3', b_3'] \subset [t_m, \infty)$ we can apply $(B.9)$ and obtain

$$\int_{a_1'}^{b_1'} \varphi(x)\, dx < \varepsilon \quad \text{and} \quad \int_{a_3'}^{b_3'} \varphi(x)\, dx < \varepsilon \,.$$

Hence,

$$P(S_n^* \in I_1) \geq \int_{a_1'}^{b_1'} \varphi(x)\, dx - \varepsilon$$

and

$$P(S_n^* \in I_3) \geq \int_{a_3'}^{b_3'} \varphi(x)\, dx - \varepsilon \,.$$

Furthermore, since $I_2 \subset [a_2', b_2'] \subset [t_1, t_m]$ we conclude by $(B.10)$ that

$$P(S_n^* \in I_2) \geq \int_{a_2'}^{b_2'} \varphi(x)\, dx - \varepsilon$$

for $n \geq N(\varepsilon)$. Hence,

$$P(a \leq S_n^* \leq b) \quad = \quad P(S_n^* \in I_1) + P(S_n^* \in I_2) + P(S_n^* \in I_3)$$

$$\geq \quad \int_{a_1'}^{b_1'} \varphi(x)\, dx - \varepsilon + \int_{a_1'}^{b_1'} \varphi(x)\, dx - \varepsilon + \int_{a_1'}^{b_1'} \varphi(x)\, dx - \varepsilon$$

$$= \quad \int_a^b \varphi(x)\, dx - 3\,\varepsilon$$

for $n \geq N(\varepsilon)$.

To prove the other inequality in $(B.7)$ observe that for $n \geq N(\varepsilon)$

$$P(S_n^* \leq t_1) \leq 2\varepsilon \quad \text{and} \quad P(S_n^* \geq t_m) \leq 2\varepsilon . \tag{B.11}$$

This is a consequence of $(B.9)$ and Proposition B.4 where in particular we have shown that

$$P(t_1 \leq S_n^* \leq t_m) \to \int_{t_1}^{t_m} \varphi(x)\, dx \geq 1 - 2\varepsilon$$

as $n \to \infty$.

As above we write $P(a \leq S_n^* \leq b)$ as

$$P(a \leq S_n^* \leq b) = P(S_n^* \in I_1) + P(S_n^* \in I_2) + P(S_n^* \in I_3)$$

and we estimate the expressions on the right-hand side.

Since $I_2 \subset [a_2', b_2'] \subset [t_1, t_m]$ we can use $(B.10)$ and obtain

$$P(S_n^* \in I_2) \leq \int_{a_2'}^{b_2'} \varphi(x)\, dx + \varepsilon$$

for $n \geq N(\varepsilon)$.

Since $I_1 \subset [a_1', b_1'] \subset (-\infty, t_1]$ we apply $(B.11)$ and infer that

$$P(S_n^* \in I_1) \leq P(S_n^* \in (-\infty, t_1]) = P(S_n^* \leq t_1) \leq 2\varepsilon$$

for $n \geq N(\varepsilon)$.

Analogously since $I_3 \subset [a_3', b_3'] \subset [t_m, \infty)$ we apply $(B.11)$ once more and obtain that

$$P(S_n^* \in I_3) \leq P(S_n^* \in [t_m, \infty)) = P(S_n^* \geq t_m) \leq 2\varepsilon$$

for $n \geq N(\varepsilon)$.

Finally we collect the above estimates and obtain

$$P(a \leq S_n^* \leq b) \leq \int_{a_2'}^{b_2'} \varphi(x)\,dx + \varepsilon + 2\varepsilon + 2\varepsilon$$

and, hence, also

$$P(a \leq S_n^* \leq b) \leq \int_{a_1'}^{b_1'} \varphi(x)\,dx + \int_{a_2'}^{b_2'} \varphi(x)\,dx + \int_{a_3'}^{b_3'} \varphi(x)\,dx + 5\varepsilon = \int_a^b \varphi(x)\,dx + 5\varepsilon$$

for $n \geq N(\varepsilon)$. $\qquad\qquad\qquad\qquad\qquad\qquad\qquad\qquad\qquad\qquad\qquad$ $\square$

B.2   Proof of the Theorem of de Moivre–Laplace

We have now collected all preliminary results needed for proving the weak convergence of S_n^* to the standard normal random variable, i.e. we show that

$$\lim_{n\to\infty} \left| P(S_n^* \leq s) - \int_{-\infty}^s \varphi(x)dx \right| = 0$$

for each $s \in \mathbb{R}$. Fix arbitrary $\varepsilon > 0$ and $s \in \mathbb{R}$. We will show that

$$\left| P(S_n^* \leq s) - \int_{-\infty}^s \varphi(x)dx \right| < \varepsilon$$

for $n \geq N(\varepsilon)$ for some sufficiently large $N(\varepsilon)$.
Choose a sequence $a_n \to -\infty$ such that

$$P(S_n^* \leq s) = P(a_n \leq S_n^* \leq s)$$

for each $n \in \mathbb{N}$. Such a sequence exists since

$$P(S_n^* \leq M) = 0$$

for $-M$ sufficiently large. Then, since

$$| P(S_n^* \leq s) - F_{\mathcal{N}}(s) | = | P(S_n^* \leq s) - P(a_n \leq S_n^* \leq s)$$

$$+ P(a_n \leq S_n^* \leq s) - (F_{\mathcal{N}}(s) - F_{\mathcal{N}}(a_n)) - F_{\mathcal{N}}(a_n) | ,$$

we obtain

$$|P(S_n^* \leq s) - F_{\mathcal{N}}(s)| \quad \leq \quad |P(S_n^* \leq s) - P(a_n \leq S_n^* \leq s)|$$

$$+ \quad |P(a_n \leq S_n^* \leq s) - (F_{\mathcal{N}}(s) - F_{\mathcal{N}}(a_n))|$$

$$+ \quad |F_{\mathcal{N}}(a_n)| \quad .$$

By the definition of a_n the first expression on the righthand side of the above inequality vanishes.

By Proposition $B.5$ we can estimate the second expression by

$$|P(a_n \leq S_n^* \leq s) - (F_{\mathcal{N}}(s) - F_{\mathcal{N}}(a_n))| \leq \frac{\varepsilon}{2}$$

for $n \geq N(\varepsilon)$ where $N(\varepsilon)$ does not depend on the particular choice of the sequence a_n.

Finally, since $a_n \to -\infty$ we obtain

$$F_{\mathcal{N}}(a_n) = \int_{-\infty}^{a_n} \varphi(x)dx \to 0$$

as $n \to \infty$. We have thus shown that

$$|P(S_n^* \leq s) - F_{\mathcal{N}}(s)| \leq \varepsilon$$

for n sufficiently large. $\square$

Bibliography

[1] C.D. Aliprantis and O. Burkinshaw: Positive Operators, Pure and Applied Mathematics Series Nr. 119, Academic Press, New York, 1985.

[2] C.D. Aliprantis, D.J. Brown and O. Burkinshaw: Existence and optimality of competitive equilibria, Springer, Heidelberg, 1990.

[3] R. Ash and C.A. Dolans-Dade: Probability and Measure Theory, 2nd ed., Academic Press, New York, 2000.

[4] S. Axler: Linear Algebra Done Right, Springer, New York, 1996.

[5] M. Baxter and A. Rennie: Financial Calculus: An Introduction to Derivative Pricing, Cambridge University Press, Cambridge, 1996.

[6] E. Behrends: Introduction to Markov Chains, Advanced Lectures in Mathematics, Vieweg, Wiesbaden, 2000.

[7] A. Bensoussan: On the theory of option pricing, Acta. Appl. Math., 2, 139–158, 1984.

[8] P. Billingsley: Probability and Measure, 3rd Ed., Wiley Series in Probability and Mathematical Statistics, John Wiley & Sons, New York, 1995.

[9] F. Black and M.S. Scholes: The Pricing of Options and Corporate Liabilities, Journal of Political Economy, 81 (3), 637–654, 1973.

[10] E. Briys and F. De Varenne: The Fisherman and the Rhinoceros: How International Finance shapes Everyday Life, John Wiley & Sons, New York, 1999.

[11] N. Cameron: Introduction to Linear and Convex Programming, Australian Mathematical Society, Lectures Series 1, Cambridge University Press, London, 1985.

[12] Y.S. Chow, H. Robbins and D. Siegmund: The Theory of Optimal Stopping, 2nd Ed., Dover, New York, 1991.

[13] K.L. Chung: Elementary Probability Theory with Stochastic Processes, Springer, New York, 1979.

[14] V. Chvátal: Linear Programming, Freeman, New York, 1983.

[15] Ph. G. Ciarlet: Introduction to Numerical Linear Algebra and Optimisation, Cambridge Texts in Applied Mathematics, Cambrige University Press, 1989.

[16] S.A. Clarck: The Valuation Problem in Arbitrage Price Theory, Journal of Mathematical Economics, 22, 463–478, 1993.

[17] J.C. Cox, S.A. Ross and M. Rubinstein: Option Pricing: A Simplified Approach, Journal of Financial Economics, 7, 229–263, 1979.

[18] J.C. Cox and M. Rubinstein: Option Markets, Prentice Hall, New Jersey, 1985.

[19] F. Delbaen and W. Schachermayer: A general version of the fundamental theorem of asset pricing, Mathematische Annalen, 300, 463–520, 1994.

[20] M.U. Dothan: Prices in Financial Markets. Oxford University Press, Oxford, 1990.

[21] D. Duffie: Dynamic Asset Pricing Theory, Second edition, Princeton University Press, Princeton, NJ, 1996.

[22] Ph.H. Dybvig and S.A. Ross: Arbitrage, in: M. Milgate, J. Eatwell and P. Newman (editors), The New Palgrave: A Dictionary of Economics; McMillan, 1987.

[23] N. El Karoui, H. Geman and J.C. Rochet: Changes of numéraire, arbitrage and option prices, J. Appl. Probab., 32, 443–458, 1995.

[24] H. Geman: L'importance de la probabilité "forward neutre" dans une approche stochastique des taux d'intérêt, Working Paper, ESSEC, 1989.

[25] P.R. Halmos: Finite-Dimensional Vector Spaces, Springer, New York, 1974.

[26] J.M. Harrison and D.M. Kreps: Martingales and Arbitrage in Multi-period Securities Markets, Journal of Economic Theory, 20, 381–408, 1979.

[27] J.M. Harrison and S.R. Pliska: Martingales and Stochastic Integrals in the Theory of Continuous Trading, Stochastic Processes and their Applications,11, 215–260, 1981.

[28] J.-B. Hiriart-Urruty and C. Lemaréchal: Fundamentals of Convex Analysis, Springer, Berlin, 2001.

[29] J. Hull: Options, Futures and Other Derivative Securities, 3rd Ed., Prentice Hall, Englewoods Cliffs, N.J., 1997

[30] J.E. Ingersoll, Jr.: Theory of Financial Decision Making, Rowman & Little-field, 1987.

[31] A. Irle: Finanzmathematik: Die Bewertung von Derivaten. Teubner, Stuttgart, 1998.

[32] F. Jamshidian: An exact bond option pricing formula, J. Finance, 44, 205–209, 1989.

[33] K. Jänich: Linear Algebra, Springer, New York,1994.

[34] R.A. Jarrow: Modelling Fixed Income Securities and Interest Rate Options, McGraw Hill, 1996.

[35] I. Karatzas and S.E. Shreve: Brownian Motion and Stochastic Calculus, Springer, New York, 1988.

[36] A.F. Karr: Probability, Springer, New York, 1993.

[37] P.E. Kopp: Martingales and Stochastic Integrals, Cambridge University Press, Cambridge, 1984.

[38] U. Krengel: Einfhrung in die Wahrscheinlichkeitstheorie, Vierte, erweiterte Auflage, Vieweg Studium, Braunschweig, Wiesbaden, 1998.

[39] D. Lamberton and B. Lapeyre: Introduction to Stochastic Calculus Applied to Finance, Chapman & Hall, London, 1996.

[40] D.P.J. Leisen: Pricing the American put option: a detailed convergence analysis for binomial models, Journal of Economic Dynamics and Control, 22, 1419–1444, 1998.

[41] S.F. LeRoy and J. Werner: Principles of Financial Economics, Cambridge University Press, Cambridge, 2000.

[42] M. Magill and M. Quinzii: Theory of Incomplete Markets, MIT Press, Cambridge, Massachusetts, 1993.

[43] R. Merton: Theory of Rational Option Pricing, Bell Journal of Economics and Management Sciences, 4(1), 141–183, 1973.

[44] R. Merton: Continuous-Time Finance, Blackwell, Oxford, 1990.

[45] M. Musiela and M. Rutkowski: Martingale Methods in Financial Modelling, Applications of Mathematics Vol. 36, Springer, New York, 1997.

[46] J. Neveu, Discrete-Parameter Martingales, North-Holland, Amsterdam, 1975.

[47] S.R. Pliska: Introduction to Mathematical Finance: Discrete Time Models, Blackwell, Oxford, 1997.

[48] M. Reimer and K. Sandmann: A discrete time approach for European and American barrier options, SFB 303, Universität Bonn, working paper B-272.

[49] T. Rolski, H. Schmidli, V. Schmidt and J. Teugels: Stochastic Processes for Insurance and Finance, Wiley Series in Probability and Statistics, John Wiley & Sons, New York, 1999.

[50] W. Rudin: Principles of Mathematical Analysis, 3rd ed., McGraw-Hill, New York, 1976.

[51] H.H. Schaefer: Banach Lattices and Positive Operators, Springer, New York, 1974.

[52] A.N. Shiryayev: Optimal Stopping Rules, Springer, NewYork, 1978.

[53] G. Strang: Linear Algebra and its Applications, 2nd ed., Academic Press, New York, 1980.

[54] H.R. Varian: The Arbitrage Principle in Financial Economics, Economic Perspectives, 1(2), 55–72, 1987.

[55] D. Williams: Probability with Martingales, Cambridge University Press, Cambridge, 1991.

[56] P. Wilmott: Derivatives: The theory and Practice of Financial Engineering, John Wiley & Sons, New York, 1998.

Index